LASER CONTROL OF CHEMICAL DYNAMICS

With Emphasis on Nonadiabatic Transition

Other World Scientific Titles by the Author

Nonadiabatic Transition: Concepts, Basic Theories and Applications
ISBN: 978-981-02-4719-5

Nonadiabatic Transition: Concepts, Basic Theories and Applications
Second Edition
ISBN: 978-981-4329-77-4

Introduction to Nonadiabatic Dynamics
ISBN: 978-981-12-0341-1

LASER CONTROL OF CHEMICAL DYNAMICS

With Emphasis on Nonadiabatic Transition

HIROKI NAKAMURA

Institute for Molecular Science, National Institutes of Natural Sciences, Okazaki, Japan

The Graduate University for Advanced Studies, Hayama, Japan

World Scientific

NEW JERSEY • LONDON • SINGAPORE • BEIJING • SHANGHAI • HONG KONG • TAIPEI • CHENNAI • TOKYO

Published by

World Scientific Publishing Co. Pte. Ltd.
5 Toh Tuck Link, Singapore 596224
USA office: 27 Warren Street, Suite 401-402, Hackensack, NJ 07601
UK office: 57 Shelton Street, Covent Garden, London WC2H 9HE

Library of Congress Cataloging-in-Publication Data
Names: Nakamura, Hiroki, author.
Title: Laser control of chemical dynamics : with emphasis on nonadiabatic transition /
Hiroki Nakamura, National Institutes of Natural Sciences, Japan,
Graduate University for Advanced Studies, Japan.
Description: New Jersey : World Scientific, [2025] | Includes bibliographical references and index.
Identifiers: LCCN 2024024229 | ISBN 9789811295812 (hardcover) |
ISBN 9789811295829 (ebook for institutions) | ISBN 9789811295836 (ebook for individuals)
Subjects: LCSH: Lasers in chemistry. | Laser manipulation (Nuclear physics) |
Charge exchange. | Molecular dynamics.
Classification: LCC QD63.L3 N35 2025 | DDC 542/.84--dc23/eng/20240629
LC record available at https://lccn.loc.gov/2024024229

British Library Cataloguing-in-Publication Data
A catalogue record for this book is available from the British Library.

For any available supplementary material, please visit
https://www.worldscientific.com/worldscibooks/10.1142/13919#t=suppl

Desk Editor: Shaun Tan Yi Jie

Typeset by Stallion Press
Email: enquiries@stallionpress.com

Preface

Controlling chemical reactions is a holy grail in chemistry. Catalysis chemistry has had an excellent long history and constitutes a strong field. Basically, catalysts change the molecular electronic states effectively. Lasers cannot modify the electronic states, on the other hand, but is rather efficient in controlling dynamics. It is true that a very strong laser can modify the electronic structures of molecules, but that strong laser can easily induce many undesirable multi-photon processes and break up the molecule eventually. Thus, it is not recommended to apply strong lasers to molecules [1]. Thanks to the recent remarkable progress in laser technology, it has become possible to design the laser pulse shape and frequency as a function of time in such a way to control molecular dynamics [1, 2]. It is possible to think about controlling the motion of quantum mechanical wave packet on a single adiabatic potential surface and nonadiabatic transitions between two adiabatic potential energy surfaces. Furthermore, lasers can move molecular electronic states up and down by the amount of photon energy. Such new states are called dressed states. This means that new potential energy surface crossings, i.e., conical intersections, are created and the nonadiabatic transitions are induced there.

The author's research group has investigated the various possibilities of laser control of chemical dynamics based on the ideas of dressed states and nonadiabatic transitions [3, 4]. This book is written as a sort of summary of such achievements. I want to express my sincere thanks to Aquisitions Assistant Director Shaun Tan Yi Jie of World Scientific Publishing Co. for this suggestion.

I am also very much indebted to many of my collaborators for their excellent accomplishments in the research of laser control of chemical dynamics. They are, in alphabetical order, Drs. J. Fujisaki, T. Ishida, A. Kondorskiy, G. Mil'nikov, K. Nagaya, S. Nanbu, H. Tamura, Y. Teranishi and S. Zou. Gratitude from the depth of my heart should also be extended to many other collaborators, especially for the developments of the Zhu-Nakamura theory of nonadiabatic transition and its applications.

Acknowledgment is also due to the following publishers for the permission of reproducing various figures of our own works: American Chemical Society, American Institute of Physics, American Physical Society, Royal Society of Chemistry and World Scientific. I have used many figures from our own papers published therein. In Chapter 8 I have used figures from my Chapter 5 of the book "Advances in Laser Physics and Technology" published by Cambridge University Press [5]. Permission to reuse was not formally obtained due to the irony of the book not being registered in PLSclear despite the publisher's instruction to submit all book permission requests through PLSclear, but I trust that the publisher would not deny me this small privilege.

Finally, I would like to express my heartfelt thanks to my wife, Suwako, for her continual support in my life.

Hiroki Nakamura
Okazaki, Japan
May 2024

References

[1] A. Bandrauk (Ed.), *Molecules in Laser Fields*, Marcel Dekker, 1994.
[2] S.A. Rice and M. Zhao, *Optical Control of Molecular Dynamics*, John Wiley & Sons, 2000.
[3] H. Nakamura, *Nonadiabatic Transition: Concepts, Basic Theories and Applications*, World Scientific, 2002 (1st edition), 2012 (2nd edition).
[4] H. Nakamura, *Introduction to Nonadiabatic Dynamics*, World Scientific, 2019.
[5] Y. Teranishi, H. Nakamura and S.H. Lin, Chapter 5 in *Advances in Laser Physics and Technology*, Cambridge University Press, 2014.

About the Author

Dr. Hiroki Nakamura is Professor Emeritus at the Institute for Molecular Science (IMS), National Institutes of Natural Sciences (NINS), Japan, where he was formerly IMS Director-General and NINS Vice-President. He is also Professor Emeritus at the Graduate University for Advanced Studies, Japan, and Honorary Professor at Xi'an Jiaotong University, China. He has authored >200 peer-reviewed papers, delivered >100 invited talks at international and major domestic conferences, wrote >60 reviews and book chapters, and published 6 books.

Contents

Chapter 1

Introduction

Recent progress in chemical dynamics and laser technology has been remarkable and control of chemical dynamics by laser is no longer an unrealizable dream. Various ideas have been proposed so far such as coherent control, adiabatic rapid passage, pump and dump, quadratic chirping, utilization of complete reflection phenomenon, and optimal control theory as explained in various reviews and books [1–12]. In this book theoretical possibilities and considerations on controlling chemical dynamics by laser are presented based on the theoretical achievements made by the author's research group.

There are three basic elements in chemical dynamics: (1) excitation and de-excitation of energy levels or pump-dump of wave packet between two adiabatic potential energy surfaces, (2) transitions between two adiabatic potential energy surfaces at their crossings, and (3) motion of quantum wave packet on an adiabatic potential energy surface. If we could control these basic processes as we desire, control of various chemical dynamics would be possible. It should be noted that there are two very important concepts: one is *dressed state* and the other is *nonadiabatic transition*. Dressed states are the states created by laser-molecule interactions. The naturally existing states are shifted up (down) by photon absorption (emission) by the amount of photon energy. Thus, the crossings of potential energy curves or surfaces are created among them artificially. This means that the transitions between those states occur efficiently. Ordinary excitation and de-excitation can be re-interpreted by this picture in the situation of a strong light field. The second important concept of nonadiabatic transition is nothing but the transition between two adiabatic potential energy curves or surfaces. The crossings between two adiabatic potential energy surfaces are called *conical intersections* [13]. As can be comprehended from the concept of dressed state, there can be two kinds of conical intersections, the first being naturally existing and the second being among the dressed states. The significance of nonadiabatic transition in chemical dynamics cannot be overlooked [5, 12]. In a sense, nonadiabatic transition is one of the very basic mechanisms of the mutability of this world. Whenever nonadiabatic transitions due to potential curve or surface crossings have to be treated, the Zhu-Nakamura theory can be usefully utilized [5, 12]. This theory is a generalization of the famous pioneering works done in 1932 by Landau, Zener and Stueckelberg [14–16].

Two of the basic elements in chemical dynamics mentioned above, (1) and (2), are nothing but nonadiabatic transitions. We have proposed the method of periodic sweeping of laser parameters to control the nonadiabatic transitions among the dressed states. This is a generalization of the linear chirping or adiabatic rapid passage [4]. The quadratic chirping is most effective and can achieve almost complete transition. For controlling the motion on a single adiabatic potential energy surface, the basic element (3) mentioned above, we have proposed the *directed momentum method* and the *semiclassical guided optimal control theory.* In the directed momentum method the momentum of wave packet motion is controlled into a desirable direction by using the coordinate dependence of dipole moment. The semiclassical guided optimal control theory has been formulated so that the multi-dimensional dynamics can be controlled with high efficiency. The theory is designed by using the Herman-Kluk-type frozen Gaussian wave packet propagation method [17]. In order to attain high efficiency the appropriate intermediate target states are set to guide the initial wave packet toward the target state. These methods can be used also for controlling the nonadiabatic transitions at naturally existing conical intersections. That is to say, the initial wave packet can be directed toward the desirable conical intersection position with desirable momentum so that the transition there can be efficient. One of the most efficient ways of controlling chemical reactions is to change the electronic structure or to change the topography of potential energy surfaces. This is actually done by catalysts in catalytic chemistry. It is true that we can change the geometry of potential energy surfaces by using a very strong laser field, but it causes many undesirable multi-photon processes and is not appropriate. The laser intensity is recommended to be less than $\sim$10 TWatt/cm^2 = 1×10^{13} Watt/cm^2 [18]. It is worthwhile to note how strong the Coulombic force in the hydrogen atom is. As is well known, the ionization potential of the hydrogen atom is

$$V_0 \equiv e^2/a_0 = 4.36 \times 10^{-18} \text{ joule} = 27.21 \text{ eV}. \tag{1.1}$$

The corresponding electric field E_0 and the laser intensity I_0 are

$$\begin{aligned} E_0 \equiv e/a_0^2 = 5.15 \times 10^9 \text{ Volt/cm}, \quad \text{and} \\ I_0 = 3.51 \times 10^{16} \text{ Watt/cm}^2. \end{aligned} \tag{1.2}$$

In addition to the above-mentioned three elementary processes, it is also worthwhile to think about the applications of the intriguing phenomenon of *complete reflection* [5, 12]. This occurs in the nonadiabatic tunneling-type potential crossing in which two diabatic potential curves cross with opposite signs of slopes (see Fig. 1.1). This phenomenon occurs at certain discrete energies higher than the bottom of the upper adiabatic potential. This cannot be complete in multi-dimensional systems, but it may survive to some extent and can be used to control some chemical dynamics and also to manifest some molecular functions [5, 12].

This book is organized as follows. In Chapters 2–4 basic background theories are presented and in Chapters 5–8 our basic ideas of controlling chemical dynamics are summarized and explained. In Chapter 2 semiclassical representation of the photon field is described. Although the photon field should be, in principle, treated as the quantized electromagnetic field, in the case of strong light field like laser it can be treated as the classical oscillatory electromagnetic field with definite amplitude and phase. In Chapter 3 the Floquet theorem and dressed state picture are introduced. Taking time average of the wave functions over the period $T(= 2\pi/\omega)$ leads to

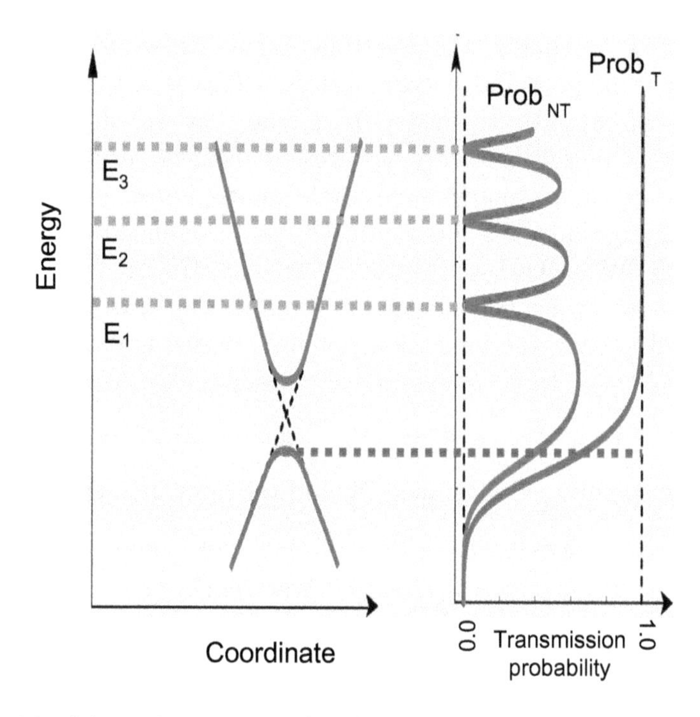

Figure 1.1: Schematic picture to show how different nonadiabatic tunneling is from ordinary tunneling. $P_T(P_{NT})$ is the transmission probability in the case of single adiabatic potential [blue curve] (coupled nonadiabatic tunneling type potential system [red curve]). P_{NT} becomes exactly zero at certain discrete energies $E_j(j = 1, 2, 3 \ldots)$. Reproduced with permission from [19].

the picture of Floquet or dressed state which presents a very useful concept for laser control. Here ω is the laser frequency. In Chapter 4 the Zhu-Nakamura theory of nonadiabatic transition is presented. This can be usefully utilized in the various control schemes explained in the following chapters. In Chapter 5 periodic sweeping (chirping) of laser parameters is explained and actual numerical applications are presented such as selective and complete excitation of energy levels and photo-dissociation of diatomic molecules with high efficiency by quadratically chirped pulses. In Chapter 6 intriguing phenomenon of complete reflection in the nonadiabatic tunneling-type potential system is introduced. At certain discrete energies $E_1, E_2, E_3, \ldots$

the transmission probability becomes exactly zero. Selective photo-dissociation of HI, CH_3SH and HOD is numerically demonstrated. In Chapter 7 semiclassical guided optimal control theory and the directed momentum method are formulated. These can be efficiently employed to move the wave packet on a single adiabatic potential energy surface. Numerical demonstrations are provided for photo-conversion of CHD to HT, selective photo-dissociation of OHCl and vibrational isomerization of HCN. In Chapter 8 it is shown that even continuous wave lasers can be utilized to control certain kinds of chemical reactions. Numerical demonstrations are given by using some model systems. Concluding remarks and future perspectives are presented in Chapter 9.

References

[1] R.D. Levine, *Molecular Reaction Dynamics*, Cambridge University Press, 2005.
[2] H. Nakamura, *J. Phys. Chem.* **A110**, 10929 (2006).
[3] P. Brumer and M. Shapiro, *Annu. Rev. Phys. Chem.* **48**, 601 (1997).
[4] S.A. Rice and M. Zhao, *Optical Control of Molecular Dynamics*, John Wiley & Sons, 2000.
[5] H. Nakamura, *Nonadiabatic Transition: Concepts, Basic Theories and Applications*, World Scientific, 2002 (1st edition), 2012 (2nd edition).
[6] A.D. Bandrauk, Y. Fujimura and R.J. Gordon, *Laser Control and Manipulation of Molecules* (ACS Symposium Series 821); American Chemical Society, Washington DC, 2002.
[7] P. Brumer and M. Shapiro, *Principles of the Quantum Control of Molecular Processes*, John Wiley & Sons, 2003.
[8] K. Yamanouchi, S.L. Chin, P. Agostini and G. Ferrante (Eds.), *Progress in Ultrafast Intense Laser Science*, Springer, 2006.
[9] V. Bonacic-Koutecky and R. Mitric, *Chemical Reviews* **105**, 11 (2005).
[10] H. Nakamura, *Adv. Chem. Phys.* **138**, 95 (2008).
[11] L.G.G. Rego, L. Santos and V.S. Batista, *Annu. Rev. Chem. Phys.* **60**, 293 (2009).
[12] H. Nakamura, *Introduction to Nonadiabatic Dynamics*, World Scientific, 2019.
[13] W. Domcke, D. Yarkony and H. Köppel, *Conical Intersections*, World Scientific, 2004.

[14] L.D. Landau, *Phys. Zts. Sov.* **2**, 46 (1932).
[15] C. Zener, *Proc. Roy. Soc.* **A137**, 696 (1932).
[16] E.C.G. Stueckerberg, *Hel. Phys. Acta.* **5**, 369 (1932).
[17] M.F. Herman, *Annu. Rev. Phys. Chem.* **45**, 83 (1994).
[18] A. Bandrauk (Ed.), *Molecules in Laser Fields*, Marcel Dekker, 1994.
[19] S. Nanbu, T. Ishida and H. Nakamura, *Chem. Sci.* **1**, 663 (2010).

Chapter 2

Semiclassical Representation of Photon Field and Laser-Matter Interaction

Since photons are quantum mechanical *particles*, in order to correctly treat the interactions of photons with atoms and molecules the electromagnetic field should be quantized and the number representation of photon field should be employed. In the case of strong field, however, it is a good approximation to treat it as the classical time-dependent sinusoidal oscillatory field [1, 2]. This is true in the case of lasers, since the photon number n and its fluctuation δn satisfy the relation,

$$n \gg \delta n \gg 1 \tag{2.1}$$

and the field is coherent with a definite phase. The photon number and phase satisfy the uncertainty relation and the uncertainty of photon number is large in the case of large photon number. In this case the phase can be defined accurately and the field can be treated as an oscillatory electric field with definite amplitude and phase.

The Schrödinger equation that describes the motion of a charged particle in the classical electromagnetic field can be derived as follows. With use of the vector potential $\mathbf{A}(\mathbf{r}, t)$ and the scalar potential $\phi(\mathbf{r})$ the Hamiltonian is given by

$$H = \frac{1}{2m}\left(\mathbf{p} - \frac{q}{c}\mathbf{A}(\mathbf{r}, t)\right)^2 - q\phi(\mathbf{r}), \tag{2.2}$$

where $q, \mathbf{p}, m$ are respectively charge, momentum and mass of the charged particle, and c is the speed of light. Under the ordinary quantization procedure, the Schrödinger equation becomes

$$i\hbar\frac{\partial}{\partial t}\psi(\mathbf{r},t) - \left[\frac{1}{2m}\left(-i\hbar\nabla - \frac{q}{c}\mathbf{A}(\mathbf{r},t)\right)^2 + V(\mathbf{r})\right]\psi(\mathbf{r},t)$$

$$\equiv i\hbar\frac{\partial}{\partial t}\psi(\mathbf{r},t) - [H_0(\mathbf{r}) + H_I(\mathbf{r},t)]\psi(\mathbf{r},t) = 0. \quad (2.3)$$

Here, H_0 and H_I represent the unperturbed Hamiltonian and the interaction of the charged particle with the field, respectively. These are explicitly expressed as follows:

$$H_0(\mathbf{r}) = -\frac{\hbar^2}{2m}\nabla^2 + V(\mathbf{r}), \quad (2.4)$$

$$V(\mathbf{r}) = -q\phi(\mathbf{r}), \quad (2.5)$$

$$H_I(\mathbf{r},t) = \frac{q}{2mc}\left[i\hbar\mathbf{A}(\mathbf{r},t)\cdot\nabla + i\hbar\nabla\cdot\mathbf{A}(\mathbf{r},t)\right] + \frac{q^2}{2mc^2}\mathbf{A}^2(\mathbf{r},t). \quad (2.6)$$

If we employ the Coulomb gauge,

$$\nabla\cdot\mathbf{A}(\mathbf{r},t) = 0, \quad (2.7)$$

the interaction Hamiltonian H_I is re-expressed as

$$H_I(\mathbf{r},t) = \frac{i\hbar q}{mc}\mathbf{A}(\mathbf{r},t)\cdot\nabla + \frac{q^2}{2mc^2}\mathbf{A}^2(\mathbf{r},t). \quad (2.8)$$

Under the Coulomb gauge condition the vector potential $\mathbf{A}(\mathbf{r},t)$ satisfies the following differential equation:

$$\left(\nabla^2 - \frac{1}{c^2}\frac{\partial^2}{\partial t^2}\right)\mathbf{A}(\mathbf{r},t) = 0. \quad (2.9)$$

Then the vector potential $\mathbf{A}(\mathbf{r},t)$ can be expressed as a superposition of plane waves,

$$\mathbf{A}(\mathbf{r},t) = \sum_k \epsilon_k\left[q_k(t)\exp(i\mathbf{k}\cdot\mathbf{r}) + q_k^*(t)\exp(-i\mathbf{k}\cdot\mathbf{r})\right], \quad (2.10)$$

where $\epsilon_{\mathbf{k}}$ represents the unit vector in the direction of polarization and $\mathbf{k}$ is the wave vector of plane wave. From the gauge condition

$$\epsilon_{\mathbf{k}}\cdot\mathbf{k} = 0 \quad (2.11)$$

holds and thus $q_k(t)$ satisfies the first-order differential equation,

$$\frac{\partial}{\partial t}q_k(t) = -\omega_k q_k(t), \tag{2.12}$$

where $\omega = ck$. In the case that the field is composed of a single mode, $q(t)$ can be simply expressed as

$$q(t) = A_0 \exp[-i(\omega t + \delta)], \tag{2.13}$$

where δ is a certain phase. Then, the vector potential $\mathbf{A}(\mathbf{r}, t)$ can be given by

$$\mathbf{A}(\mathbf{r}, t) = \epsilon A_0 \cos(\mathbf{k} \cdot \mathbf{r} - \omega t - \delta). \tag{2.14}$$

Ordinarily, the wave length $(2\pi/k)$ of the electromagnetic wave is much longer than the size of an atom or molecule, namely k is small; thus $\mathbf{k} \cdot \mathbf{r} \ll 1$ holds and $\exp(i\mathbf{k} \cdot \mathbf{r}) \sim 1$ is satisfied. This is called the dipole moment approximation. Finally, the vector potential $\mathbf{A}(\mathbf{r}, t)$ does not depend on the spatial coordinate $\mathbf{r}$ and is expressed as

$$\mathbf{A}(t) = \epsilon A_0 \cos(\omega t + \delta). \tag{2.15}$$

If the laser is pulse-shaped, the intensity depends on time, namely,

$$\mathbf{A}(t) = \epsilon A_0(t) \cos(\omega t + \delta). \tag{2.16}$$

Under the dipole moment approximation, the interaction Hamiltonian H_I is given by Eq. (2.8) with use of $\mathbf{A}(t)$ given by Eq. (2.16). Using the transformation

$$\psi(\mathbf{r}, t) = \exp\left(i\frac{q}{\hbar c}\mathbf{r} \cdot \mathbf{A}(t)\right) \Psi(\mathbf{r}, t) \tag{2.17}$$

and introducing the electric field

$$\mathbf{E}(t) = -\frac{1}{c}\frac{\partial}{\partial t}\mathbf{A}(t), \tag{2.18}$$

we have from Eq. (2.3)

$$i\hbar\frac{\partial}{\partial t}\Psi(\mathbf{r}, t) = \left[-\frac{\hbar^2}{2m}\nabla^2 + V(\mathbf{r}) - q\mathbf{r} \cdot \mathbf{E}(t)\right] \Psi(\mathbf{r}, t). \tag{2.19}$$

The interaction Hamiltonian H_I becomes

$$H_I(t) = -q\mathbf{r} \cdot \mathbf{E}(t). \tag{2.20}$$

This represents the interaction between dipole moment and electric field. In the case of many-particle systems, it is good enough to simply take a summation over particles.

In the actual applications the basic equations to be dealt with are quite different for atoms and molecules. In the case of atoms the transitions among the energy levels are the basic subject. In the case of molecules, on the other hand, we have to deal with the molecular motions.

In the first case the total wave function $\Psi(\mathbf{r}, t)$ is expanded in terms of the wave functions $\{\phi_j(\mathbf{r})\}$ of the unperturbed Hamiltonian H_0,

$$\Psi(\mathbf{r}, t) = \sum_j c_j(t)\phi_j(\mathbf{r}). \tag{2.21}$$

The coefficients $\{c_j(t)\}$ satisfy the coupled differential equations,

$$i\hbar\frac{d}{dt}\begin{pmatrix} c_1(t) \\ c_2(t) \\ \cdot \\ \cdot \\ \cdot \end{pmatrix} = \begin{pmatrix} E_1 & H_{12}(t) & \cdot & \cdot & \cdot \\ H_{21}(t) & E_2 & \cdot & \cdot & \cdot \\ \cdot & \cdot & \cdot & \cdot & \cdot \\ \cdot & \cdot & \cdot & \cdot & \cdot \\ \cdot & \cdot & \cdot & \cdot & \cdot \end{pmatrix}\begin{pmatrix} c_1(t) \\ c_2(t) \\ \cdot \\ \cdot \\ \cdot \end{pmatrix}, \tag{2.22}$$

where

$$E_j = \langle\phi_j|H_0|\phi_j\rangle_{\mathbf{r}}, \quad H_{jk}(t) = -\mu_{jk}\cdot\mathbf{E}(t), \quad \mu_{jk} = \langle\phi_j|\mathbf{r}|\phi_k\rangle_{\mathbf{r}}. \tag{2.23}$$

Thus, the problem of transitions among energy levels becomes to solve the above coupled differential equations under a given initial condition.

In the case of laser control of molecular dynamic processes, the nuclear coordinates $\mathbf{R}$ in addition to the electron coordinates $\mathbf{r}$ should be taken into account. The Schrödinger equation to be solved is given by

$$i\hbar\frac{\partial}{\partial t}\Psi(\mathbf{r}, \mathbf{R}, t) = \big[H_0(\mathbf{r}, \mathbf{R}) + H_I(\mathbf{r}, \mathbf{R}, t)\big]\Psi(\mathbf{r}, \mathbf{R}, t), \tag{2.24}$$

where

$$H_0(\mathbf{r}, \mathbf{R}) = T_N + H_{el}(\mathbf{r}; \mathbf{R}), \quad H_I(\mathbf{r}, \mathbf{R}, t) = -\mu(\mathbf{r}, \mathbf{R})\cdot\mathbf{E}(t). \tag{2.25}$$

Here T_N represents the kinetic energy of relative nuclear motion and H_{el} is the molecular electronic Hamiltonian at fixed nuclear configuration. As usual the total wave function $\Psi(\mathbf{r}, \mathbf{R}, t)$ is expanded in terms of the adiabatic electronic states,

$$\Psi(\mathbf{r}, \mathbf{R}, t) = \sum_n \chi_n^{(a)}(\mathbf{R}, t)\phi_n^{(a)}(\mathbf{r}; \mathbf{R}), \tag{2.26}$$

where

$$H_{el}(\mathbf{r}; \mathbf{R})\phi_n^{(a)}(\mathbf{r}; \mathbf{R}) = E_n(R)\phi_n^{(a)}(\mathbf{r}; \mathbf{R}). \tag{2.27}$$

Inserting this expansion into Eq. (2.24), we have the following coupled differential equations for $\chi_n^{(a)}$ as

$$i\hbar\frac{\partial}{\partial t}\begin{pmatrix} \chi_1^{(a)}(\mathbf{R}, t) \\ \chi_2^{(a)}(\mathbf{R}, t) \\ \cdot \\ \cdot \\ \cdot \end{pmatrix} = \left\{ T_N\mathbf{I} + \begin{pmatrix} E_1(\mathbf{R}) & H_{12}^{(a)}(\mathbf{R}, t) & \cdot & \cdot & \cdot \\ H_{21}^{(a)}(\mathbf{R}, t) & E_2(\mathbf{R}) & \cdot & \cdot & \cdot \\ \cdot & \cdot & \cdot & \cdot & \cdot \\ \cdot & \cdot & \cdot & \cdot & \cdot \\ \cdot & \cdot & \cdot & \cdot & \cdot \end{pmatrix} \right\}$$
$$\times \begin{pmatrix} \chi_1^{(a)}(\mathbf{R}, t) \\ \chi_2^{(a)}(\mathbf{R}, t) \\ \cdot \\ \cdot \\ \cdot \end{pmatrix}, \tag{2.28}$$

where

$$H_{jk}^{(a)}(\mathbf{R}, t) = \Lambda_{jk}(\mathbf{R}) - \mu_{jk}^{(a)}(\mathbf{R}) \cdot \mathbf{E}(\mathbf{t}), \tag{2.29}$$

$$\mu_{jk}^{(a)}(\mathbf{R}) = \langle\phi_j^{(a)}|\mu(\mathbf{r}, \mathbf{R})|\phi_k^{(a)}\rangle_{\mathbf{r}}, \tag{2.30}$$

$$\Lambda_{jk}(\mathbf{R}) = -\sum_\alpha \frac{\hbar^2}{m_\alpha}\left(A_{jk}^{(a)}\frac{\partial}{\partial \mathbf{R}_\alpha} + \frac{1}{2}B_{jk}^{(a)}\right), \tag{2.31}$$

$$A_{jk}^{(a)} = \left\langle \phi_j^{(a)} \middle| \frac{\partial}{\partial \mathbf{R}_\alpha} \middle| \phi_k^{(a)} \right\rangle_{\mathbf{r}}, \quad B_{jk}^{(a)} = \left\langle \phi_j^{(a)} \middle| \frac{\partial^2}{\partial \mathbf{R}_\alpha^2} \middle| \phi_k^{(a)} \right\rangle_{\mathbf{r}}. \tag{2.32}$$

Here $\mathbf{I}$ is the unit matrix, Λ_{jk} represents the nonadiabatic couplings, and $A_{jk}^{(a)}$ represents the coupling between the two adiabatic states

E_j and E_k in the case of no laser field. On the other hand, $B_{jk}^{(a)}$ is zero for $j \neq k$ and gives a correction to the adiabatic potential. When the potential surfaces $E_j(\mathbf{R})$ and $E_k(\mathbf{R})$ have a conical intersection, the nonadiabatic transition between the two states occurs, which can be treated by the Zhu-Nakamura theory explained in Chapter 4. Another important transition is that induced by laser between the Floquet states as explained in the next chapter.

In concluding this chapter, the units to represent the strength of laser and the conversions among the commonly used units, Watt/cm^2, Volt/cm and a.u., are provided. In terms of the laser strength I, the electric field strength E in Volt/cm is given by

$$E = \sqrt{\frac{8\pi}{c} I}. \tag{2.33}$$

The conversion between a.u. and Watt/cm^2 is given by

$$I(\text{a.u.}) = I(\text{Watt/cm}^2) \times 1.55256 \times 10^{-16} \tag{2.34}$$

and

$$E(\text{a.u.}) = \sqrt{I(\text{Watt/cm}^2)} \times 5.335 \times 10^{-9}. \tag{2.35}$$

References

[1] S.I. Chu, *Advances in Multiphoton Processes and Spectroscopy, Vol. 2*, World Scientific, 1986.

[2] C. Cohen-Tannoudji, J. Dupon-Roc and G. Grynberg, *Atom-Photon Interactions*, John Wiley & Sons, 1992.

Chapter 3

Floquet Theorem and Dressed State Picture

First, let us consider an atom in a stationary laser field. The Schrödinger equation to be solved is

$$\left[i\hbar\frac{\partial}{\partial t} - H(\mathbf{r},t)\right]\Psi(\mathbf{r},t) \equiv \left[i\hbar\frac{\partial}{\partial t} - \big(H_0(\mathbf{r}) + H_I(\mathbf{r},t)\big)\right]\Psi(\mathbf{r},t) = 0 \tag{3.1}$$

with

$$H_I(\mathbf{r},t) = -\mu(\mathbf{r})\cdot\mathbf{E}_0\cos(\omega t). \tag{3.2}$$

Since the total Hamiltonian satisfies the periodicity,

$$H(\mathbf{r},t+T) = H(\mathbf{r},t) \quad \text{with } T = \frac{2\pi}{\omega}, \tag{3.3}$$

the total wave function $\Psi(\mathbf{r},t)$ can be expressed as

$$\Psi(\mathbf{r},t) = \exp(-i\epsilon t/\hbar)\Phi(\mathbf{r},t) \tag{3.4}$$

with

$$\Phi(\mathbf{r},t+T) = \Phi(\mathbf{r},t), \tag{3.5}$$

where ϵ is called Floquet quasi-energy. Inserting Eq. (3.4) into Eq. (3.1), we have

$$\left[H(\mathbf{r},t) - i\hbar\frac{\partial}{\partial t}\right]\Phi_q(\mathbf{r},t) = \epsilon_q\Phi(\mathbf{r},t). \tag{3.6}$$

The eigenfunction $\Phi_q(\mathbf{r}, t)$ satisfies the orthonormal condition,

$$\frac{1}{T}\int_0^T dt \langle \Phi_p | \Phi_q \rangle_{\mathbf{r}} = \delta_{pq} \tag{3.7}$$

and can be expanded in terms of the eigenfunctions $\phi_\beta(\mathbf{r})$ of $H_0(\mathbf{r})$ as

$$\Phi_q(\mathbf{r}, t) = \sum_m \sum_\beta C_{m\beta}^{(q)} \phi_\beta(\mathbf{r}) \exp(-im\omega t), \tag{3.8}$$

where

$$H_0(\mathbf{r})\phi_\beta(\mathbf{r}) = E_\beta \phi_\beta(\mathbf{r}). \tag{3.9}$$

Multiplying $\exp(in\omega t)\phi_\alpha^*(\mathbf{r})$ from the left and integrating over $\mathbf{r}$ and carrying out the time averaging over the period T, the secular equation with respect to the Floquet energy ϵ_q can be obtained as

$$\begin{aligned}\sum_m \sum_\beta H_{\alpha n,\beta m}^F C_{m\beta}^{(q)} &\equiv \sum_m \sum_\beta \left[H_{\alpha\beta}^{[n-m]} - m\hbar\omega \delta_{nm} \delta_{\alpha\beta} \right] C_{m\beta}^{(q)} \\ &= \epsilon_q C_{n\alpha}(q),\end{aligned} \tag{3.10}$$

where

$$H_{\alpha\beta}^{[n-m]} = \langle \phi_\alpha | H^{[k]} | \phi_\beta \rangle_{\mathbf{r}} \tag{3.11}$$

and

$$H^{[k]} = \frac{1}{T}\int_0^T dt \exp(ik\omega t)[H_0(\mathbf{r}) + H_I(\mathbf{r}, t)]. \tag{3.12}$$

In the dipole moment approximation $H^{[k]}$ is non-zero only for $k = 0$ and ± 1. The matrix elements of the Floquet Hamiltonian are finally given by

$$H_{\alpha,\beta}^F = \delta_{mn}\delta_{\alpha\beta}(E_\beta - m\hbar\omega) - (\delta_{m,n+1} + \delta_{m,n-1})\mu_{\alpha\beta} \cdot \mathbf{E}_0. \tag{3.13}$$

The Floquet matrix H^F is explicitly written as

$$H^F = \begin{pmatrix} \cdots & \cdot & \cdot & \cdot & \cdot & \cdots \\ \cdots & \cdot & \cdot & \cdot & \cdot & \cdots \\ \cdots & \cdot & \cdot & \cdot & \cdot & \cdots \\ \cdots & E_1 & 0 & 0 & -\mu_{12}\cdot\mathbf{E}_0/2 & \cdots \\ \cdots & 0 & E_2 & -\mu_{12}\cdot\mathbf{E}_0/2 & 0 & \cdots \\ \cdots & 0 & -\mu_{12}\cdot\mathbf{E}_0/2 & E_1+\hbar\omega & 0 & \cdots \\ \cdots & -\mu_{12}\cdot\mathbf{E}_0/2 & 0 & 0 & E_2+\hbar\omega & \cdots \\ \cdots & \cdot & \cdot & \cdot & \cdot & \cdots \\ \cdots & \cdot & \cdot & \cdot & \cdot & \cdots \\ \cdots & \cdot & \cdot & \cdot & \cdot & \cdots \end{pmatrix}. \tag{3.14}$$

This matrix makes actually an infinite chain, since an infinite number of photon absorption and emission are possible in principle. Besides, in the case of multi-level problem instead of the two-level (E_1, E_2) considered here, this matrix becomes a nested super-matrix. In many cases in reality, however, one-photon processes in two-state systems are important. In that case the Floquet Hamiltonian is simply given by

$$H^F = \begin{pmatrix} E_1+\hbar\omega & -\frac{\mu_{12}\cdot\mathbf{E}_0}{2} \\ -\frac{\mu_{12}\cdot\mathbf{E}_0}{2} & E_2 \end{pmatrix}. \tag{3.15}$$

As is seen from Eqs. (3.14) and (3.15), the diagonal elements have the forms of $E_j \pm n\hbar\omega$ $(n = 0, 1, 2, \ldots)$. The original energy E_j moves up or down by the photon energy $n\hbar\omega$. These states are called *dressed states*. The above content makes the Floquet theorem and the representation is called *dressed state representation* or *Floquet state representation* [1]. As is clearly understood from Eq. (3.15), the state $E_1 + \hbar\omega$ moves up when the laser frequency ω increases as a function of time and crosses the state E_2 when $E_1 < E_2$. If we diagonalize this matrix, the adiabatic states can be obtained as a function of time which take into account the

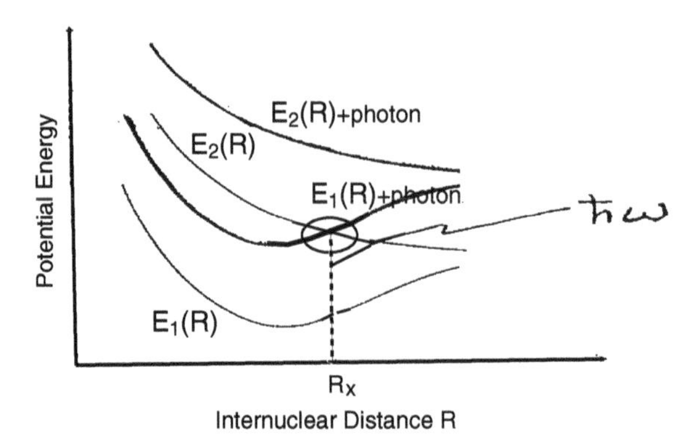

Figure 3.1: Schematic picture of dressed states of molecules.

laser-matter interactions in the infinite order and the two adiabatic states avoid crossing. This representation is very useful and convenient to deal with nonadiabatic processes in a laser field. Exactly the same treatment is possible when we consider the vibrational energy levels of molecules instead of atomic energy levels.

In the case of electronic states of molecules the energy levels become the R-dependent potential energy curves and the dressed state picture looks like that in Fig. 3.1. There is a curve crossing between $E_2(R)$ and the dressed state $E_1(R) + \hbar\omega$ as a function of R. If we diagonalize this representation, this crossing becomes an avoided crossing in the same way as in the time-dependent case and the Zhu-Nakamura theory of nonadiabatic transition can be applied (see Chapter 4) [2, 3]. Without the laser field the transition between $E_1(R)$ and $E_2(R)$ seldom occurs, but now the transition between $E_1(R) + \hbar\omega$ and $E_2(R)$ occurs efficiently. This means that the laser field induces the new transition. This is true in the ordinary photo-absorption process. However, in the ordinary case the photon field is weak and the conventional perturbation theory is good enough. When the laser field is strong, the energy separation of the avoided crossing becomes significant and the usage of the nonadiabatic transition theory becomes unavoidable.

When the laser frequency $\omega(t)$ and the intensity $E_0(t)$ depend on time slowly compared with the time variation of $\exp(i\omega t)$, the above formulation should be modified as given below. For simplicity, let us consider the case of an atom in the field. The starting Schrödinger equation becomes

$$i\hbar\frac{\partial}{\partial t}\Psi(\mathbf{r},t) = [H_0 + V(\mathbf{r},t)]\Psi(\mathbf{r},t) \tag{3.16}$$

with

$$V(\mathbf{r},t) = -\mu(\mathbf{r})E(t)\cos[\Phi(t)] \quad \text{and} \quad \Phi(t) = \int^t \omega(t)dt. \tag{3.17}$$

The total wave function $\Psi(\mathbf{r},t)$ can be expanded as before,

$$\Psi(\mathbf{r},t) = \sum_{q,n} C_{q,n}(t)\Phi_{q,n}(\mathbf{r},t), \tag{3.18}$$

where

$$\Phi_{q,n}(\mathbf{r},t) = \exp[-iq\Phi(t)]\phi_n(\mathbf{r}) \tag{3.19}$$

and

$$H_0\phi_n(\mathbf{r}) = \epsilon_n\phi_n(\mathbf{r}) \quad \text{and} \quad \langle\phi_n|\phi_m\rangle = \delta_{nm}. \tag{3.20}$$

Here q and n designate the photon number and the atomic state, respectively. Inserting this expansion into Eq. (3.16) and taking an inner product with $\Phi_{q,n}$ over $\mathbf{r}$, we have

$$\begin{aligned}
i\hbar\sum_{p,m}\frac{\partial}{\partial t}C_{p,m}(t)\exp[i(q-p)\Phi(t)] + \sum_{p} p\hbar\omega(t)C_{p,n}(t)\exp[i(q-p)\Phi(t)] \\
= \epsilon_n\sum_{p} C_{p,n}(t)\exp[i(q-p)\Phi(t)] \\
+ \sum_{p,m}\langle\phi_n|V|\phi_m\rangle\exp[i(p-q)\Phi(t)]C_{p,m}(t).
\end{aligned} \tag{3.21}$$

Here the following time average of both sides of this equation is taken:

$$\frac{\omega(t)}{2\pi}\int_{t-\pi/\omega(t)}^{t+\pi/\omega(t)} \cdots\, dt. \tag{3.22}$$

It is assumed that the time dependencies of $F(t) = E(t), \omega(t), C_{q,n}(t)$ and $\partial C_{q,n}(t)/\partial t$ are slow compared with that of $\exp[iq\Phi(t)]$ and the following relation holds:

$$\frac{\omega(t)}{2\pi}\int_{t-\pi/\omega(t)}^{t+\pi/\omega(t)} F(t)\exp[iq\Phi(t)]dt \sim \delta_{q0}F(t). \tag{3.23}$$

Then we have

$$\begin{aligned} i\hbar\frac{d}{dt}C_{q,n}(t) &= \epsilon_n C_{q,n}(t) \\ &+ \frac{\omega(t)}{2\pi}\int_{t-\pi/\omega(t)}^{t+\pi/\omega(t)} \sum_{p,m}\langle\phi_n|V|\phi_m\rangle \exp[i(q-p)\Phi(t)]dt - q\hbar\omega(t)C_{q,n}(t). \end{aligned} \tag{3.24}$$

Since the dipole moment approximation implies

$$\begin{aligned} &\frac{\omega(t)}{2\pi}\int_{t-\pi/\omega(t)}^{t+\pi/\omega(t)} \langle\phi_n|V|\phi_m\rangle \exp[i(q-p)\Phi(t)]dt \\ &\quad = \frac{1}{2}\langle\phi_n|\mu|\phi_m\rangle E(t)[\delta_{q+1,p} + \delta_{q-1,p}], \end{aligned} \tag{3.25}$$

the coupled equations to be solved are finally given by

$$\begin{aligned} i\hbar\frac{d}{dt}C_{q,n}(t) &= \epsilon_n C_{q,n}(t) - q\hbar\omega(t)C_{q,n}(t) \\ &\quad + \frac{1}{2}\sum_{m,q}\langle\phi_n|\mu|\phi_m\rangle E(t)[\delta_{q+1,p} + \delta_{q-1,p}]. \end{aligned} \tag{3.26}$$

These coupled equations provide the basic equations which describe the transitions among atomic states due to the time dependencies of the laser frequency $\omega(t)$ and the intensity $E(t)$ in the representation of dressed *diabatic* state representation. By transforming into the dressed *adiabatic* state representation, or the Floquet adiabatic state representation, the time-dependent nonadiabatic transition theories can be applied. In the case of molecular electronic states, the nuclear coordinates $\mathbf{R}$ appear as the independent variables in addition to time, as mentioned before.

References

[1] S.I. Chu, *Advances in Multiphoton Processes and Spectroscopy, Vol. 2*, World Scientific, 1986.
[2] H. Nakamura, *Nonadiabatic Transition: Concepts, Basic Theories and Applications*, World Scientific, 2002 (1st edition), 2012 (2nd edition).
[3] H. Nakamura, *Introduction to Nonadiabatic Dynamics*, World Scientific, 2019.

Chapter 4

Zhu-Nakamura Theory of Nonadiabatic Transition

The summary of the whole set of the Zhu-Nakamura formulas are provided here for both classically allowed and forbidden cases. The two types of potential systems, namely, Landau-Zener type and nonadiabatic tunneling type, should be considered separately (see Figs. 4.1 and 4.2). The Zhu-Nakamura theory has been developed since 1992 and proved to be useful for clarifying various chemical dynamics [1, 2]. Furthermore, accurate methods have been recently proposed to treat classically forbidden dynamics in multi-dimensional space [3]. This is presented in Section 4.4 below. The formulas presented in Sections 4.1–4.3 are basically valid only for the one-dimensional systems. The time-dependent version can be easily obtained from the time-independent ones and is given in the final section. In this case only the Landau-Zener type is relevant.

4.1 Basic Parameters

The most basic parameters are a^2 and b^2 given below (see Figs. 4.1 and 4.2).

$$a^2 = \frac{f(f_1 - f_2)}{8\alpha^3} \quad \text{and} \quad b^2 = \frac{\epsilon(f_1 - f_2)}{2\alpha f} \tag{4.1}$$

with

$$\begin{aligned} \epsilon &= \frac{2m}{\hbar^2}(E - E_X), \quad f_j = \frac{2m}{\hbar^2}F_j \ (j = 1, 2), \quad \alpha = \frac{2mA}{\hbar^2}, \\ \xi &= \frac{2\alpha k}{f}, \quad f = (f_1|f_2|)^{1/2}, \end{aligned} \tag{4.2}$$

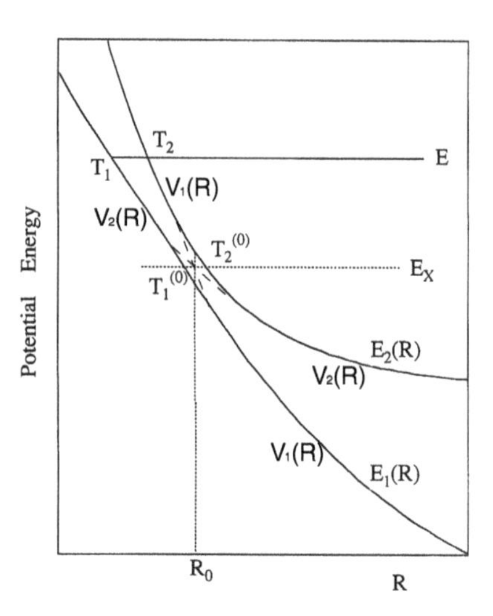

Figure 4.1: Landau-Zener-type potential energy curve crossing. $V_j(R)(j = 1, 2)$ $[E_j(R)(j = 1, 2)]$ are diabatic [adiabatic] states.

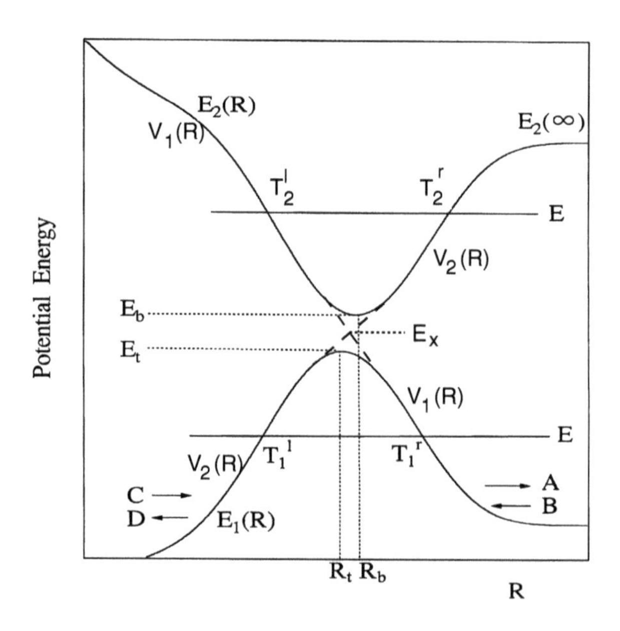

Figure 4.2: Nonadiabatic tunneling-type potential energy curve crossing. $V_j(R)(j = 1, 2)$ $[E_j(R)(j = 1, 2)]$ are diabatic [adiabatic] states.

where F_j is the slope of the diabatic potentials at the crossing point. These are valid in the case of *diabatic representation.* Generally speaking, the *adiabatic representation* is better to be used in the actual calculations. In the case of adiabatic state representation, the following expressions of a^2 and b^2 are recommended. In the case of Landau-Zener type of potential system a^2 and b^2 are defined as follows [4]:

$$a^2(\text{adia}) = \sqrt{d^2 - 1}\frac{\hbar^2}{m(T_2^{(0)} - T_1^{(0)})^2[E_2(R_0) - E_1(R_0)]} \tag{4.3}$$

and

$$b^2(\text{adia}) = \sqrt{d^2 - 1}\frac{E - [E_2(R_0) + E_1(R_0)]/2}{[E_2(R_0) - E_1(R_0)]/2} \tag{4.4}$$

with

$$d^2 = \frac{[E_2(T_1^{(0)}) - E_1(T_1^{(0)})][E_2(T_2^{(0)}) - E_1(T_2^{(0)})]}{[E_2(R_0) - E_1(R_0)]^2}, \tag{4.5}$$

where the reference position R_0 and the crossing energy E_X (see Fig. 4.1) are newly introduced as follows [4, 5]:

$$[E_2(R) - E_1(R)]_{R=R_0} = \text{minimum} \tag{4.6}$$

and

$$E_X = \frac{1}{2}[E_2(R_0) + E_1(R_0)]. \tag{4.7}$$

In the case of nonadiabatic tunneling type of potential system a^2 and b^2 are given as follows [4]:

$$a^2(\text{adia}) = \frac{\hbar^2}{m}\frac{1 - \gamma^2}{(R_b - R_t)^2(E_b - E_t)} \tag{4.8}$$

and

$$b^2(\text{adia}) = \frac{E - (E_b + E_t)/2}{(E_b - E_t)/2} \tag{4.9}$$

with

$$\gamma = \frac{E_b - E_t}{E_2(\frac{R_b+R_t}{2}) - E_1(\frac{R_b+R_t}{2})} \quad (0 < \gamma \leq 1). \tag{4.10}$$

In the limiting case of $R_b = R_t(\gamma \to 1)$, the following formula should be used for $a^2(\text{adia})$ instead of Eq. (4.8):

$$a^2(\text{adia}) = \frac{\hbar^2}{4m}\frac{E_2'' - E_1''}{(E_b - E_t)^2} \tag{4.11}$$

where

$$E_1'' = \left[\frac{d^2E_1(x)}{dx^2}\right]_{x=x_t} \quad \text{and} \quad E_2'' = \left[\frac{d^2E_2(x)}{dx^2}\right]_{x=x_b}. \tag{4.12}$$

Another important physical quantity, i.e., scattering matrix, in the adiabatic representation is defined as

$$S_{mn}^{(a)} = S_{mn}^{\mathrm{R(a)}} e^{i(\eta_m^{(a)} + \eta_n^{(a)})}, \tag{4.13}$$

where $\eta_m^{(a)}$ and $\eta_n^{(a)}$ are the elastic scattering phase shifts along adiabatic potentials and $S^{R(a)}$ is called reduced scattering matrix. The reduced scattering matrix in the Landau-Zener type of potential system is given by

$$S_{\mathrm{LZ}}^{\mathrm{R}(a)} = \begin{pmatrix} (1 + U_1U_2)e^{-2i\sigma_{ZN}} & -U_2 \\ -U_2 & (1 - U_1^*U_2)e^{2i\sigma_{ZN}} \end{pmatrix}, \tag{4.14}$$

where

$$U_2 = \frac{2i\mathrm{Im}U_1}{1 + |U_1|^2}. \tag{4.15}$$

In the case of nonadiabatic tunneling type of potential system the reduced scattering matrix is given by

$$S_{\mathrm{NT}}^{\mathrm{R}(a)} = \frac{1}{1 + U_1U_2} \begin{pmatrix} e^{i\Delta_{11}} & U_2e^{i\Delta_{12}} \\ U_2e^{i\Delta_{12}} & e^{i\Delta_{22}} \end{pmatrix}, \tag{4.16}$$

where

$$U_2 = \frac{2i(\mathrm{Im}U_1)^2}{|U_1|^2 - 1}. \tag{4.17}$$

The parameters $U_j(a^2, b^2)$ are known as Stokes constants. The explicit expressions of the Stokes constants and the phases σ_{ZN} and Δ_{ij} are given in later sections.

4.2 Classically Allowed Transition

4.2.1 *Landau-Zener Type* ($E \geq E_2(R_0)$)

The Stokes constant U_1 is given by

$$U_1 = \sqrt{\frac{1}{p_{ZN}} - 1}\,\exp(i\psi_{ZN}). \tag{4.18}$$

The nonadiabatic transition probability p_{ZN} for one passage of the crossing point is given by

$$p_{\rm ZN} = \exp\left[-\frac{\pi}{4a|b|}\left(\frac{2}{1+\sqrt{1+b^{-4}(0.4a^2+0.7)}}\right)^{1/2}\right]. \tag{4.19}$$

It should be noted that the Landau-Zener formula is simply given by

$$p_{\rm LZ} = \exp\left[-\frac{\pi}{4a|b|}\right], \tag{4.20}$$

which goes to zero as $|b| \to 0$. The correction factor $0.4a^2 + 0.7$ is originally unity and later confirmed to be slightly better than unity [6].

When the phases induced by nonadiabatic transition are needed, the following transition matrix I_X at the avoided crossing point R_0 should be used:

$$\begin{pmatrix} C \\ D \end{pmatrix} = I_X \begin{pmatrix} A \\ B \end{pmatrix}, \tag{4.21}$$

where A and B (C and D) are the coefficients of the wave functions at $R = R_0 + 0$ ($R = R_0 - 0$). This matrix gives the transition amplitude at the crossing point and explicitly defined as

$$I_X = \begin{pmatrix} \sqrt{1-p_{\rm ZN}}\,e^{i(\psi_{\rm ZN}-\sigma_{\rm ZN})} & -\sqrt{p_{\rm ZN}}\,e^{i\sigma_0^{\rm ZN}} \\ \sqrt{p_{\rm ZN}}\,e^{-i\sigma_0^{\rm ZN}} & \sqrt{1-p_{\rm ZN}}\,e^{-i(\psi_{\rm ZN}-\sigma_{\rm ZN})} \end{pmatrix}, \tag{4.22}$$

where

$$\psi_{\rm ZN} = \sigma_{\rm ZN} + \phi_{\rm S} \tag{4.23}$$

with

$$\phi_{\rm S} = -\frac{\delta_\psi}{\pi} + \frac{\delta_\psi}{\pi}\ln\left(\frac{\delta_\psi}{\pi}\right) - \arg\Gamma\left(i\frac{\delta_\psi}{\pi}\right) - \frac{\pi}{4}. \tag{4.24}$$

The various parameters are defined as follows:

$$\delta_\psi = \delta_{\rm ZN}\left(1 + \frac{5a^{1/2}}{a^{1/2}+0.8}10^{-\sigma_{\rm ZN}}\right) \tag{4.25}$$

and

$$\sigma_0^{\rm ZN} + i\delta_0^{\rm ZN} \equiv \int_{R_0}^{R_*}[K_1(R) - K_2(R)]dR \simeq \frac{\sqrt{2}\pi}{4a}\frac{F_-^C + iF_+^C}{F_+^2 + F_-^2}, \tag{4.26}$$

where

$$K_j(R) = \sqrt{\frac{2\mu}{\hbar^2}(E - E_j(R))} \quad (j = 1, 2), \tag{4.27}$$

$$F_\pm = \sqrt{\sqrt{(b^2+\gamma_1)^2+\gamma_2} \pm (b^2+\gamma_1)} + \sqrt{\sqrt{(b^2-\gamma_1)^2+\gamma_2} \pm (b^2-\gamma_1)}, \tag{4.28}$$

$$F_+^C = F_+\left(b^2 \longrightarrow \left[b^2 - \frac{0.16b_x}{\sqrt{b^4+1}}\right]\right), \tag{4.29}$$

$$F_-^C = F_-\left(\gamma_2 \longrightarrow \frac{0.45\sqrt{d^2}}{1+1.5e^{2.2b_x|b_x|^{0.57}}}\right), \tag{4.30}$$

$$b_x = b^2 - 0.9553, \quad \gamma_1 = 0.9\sqrt{d^2-1} \quad \text{and} \quad \gamma_2 = 7\sqrt{d^2}/16. \tag{4.31}$$

The correction factor in Eq. (4.25) is introduced empirically [6] and the right-hand expressions in Eq. (4.26) are derived by using the linear potential model and 3-point quadrature [5]. The parameters $\sigma_{\rm ZN}$ and $\delta_{\rm ZN}$ are given as follows:

$$\sigma_{\rm ZN} = \int_{T_1}^{R_0} K_1(R)dR - \int_{T_2}^{R_0} K_2(R)dR + \sigma_0^{\rm ZN} \tag{4.32}$$

and

$$\delta_{\rm ZN} = \delta_0^{\rm ZN}. \tag{4.33}$$

4.2.2 *Nonadiabatic Tunneling Type ($E \geq E_b$)*

The Stokes constant U_1 is given by

$$U_1 = i\sqrt{1 - p_{ZN}} \exp[i(\sigma_{ZN} - \bar{\phi}_S)]. \tag{4.34}$$

The transition probability for one passage of the crossing point is given as

$$p_{\text{ZN}} = \exp\left[-\frac{\pi}{4ab}\left(\frac{2}{1 + \sqrt{1 - b^{-4}(0.72 - 0.62a^{1.43})}}\right)^{1/2}\right]. \tag{4.35}$$

The correction factor $0.72 - 0.6a^{1.43}$ is originally unity and later confirmed to be slightly better than unity [7]. When $E \leq E_2(\infty)$, the overall transmission probability from left to right or vice versa is the physically meaningful quantity and is given by

$$P_{12} = \frac{4\cos^2(\psi_{\text{ZN}})}{4\cos^2(\psi_{\text{ZN}}) + (p_{\text{ZN}})^2/(1 - p_{\text{ZN}})} \tag{4.36}$$

with

$$\psi_{ZN} = \sigma_{ZN} - \bar{\phi}_S \quad \text{and} \quad \sigma_{\text{ZN}} = \int_{T_2^l}^{T_2^r} K_2(R)dR, \tag{4.37}$$

where $K_2(R)$ is defined by Eq. (4.27) and

$$\bar{\phi}_{\text{S}} = \phi_{\text{S}} + h_1, \tag{4.38}$$

$$h_1 = \frac{0.23a^{1/2}}{a^{1/2} + 0.75} 40^{-\sigma_{\text{ZN}}}, \tag{4.39}$$

$$\phi_{\text{S}} = -\frac{\delta_{\text{ZN}}}{\pi} + \frac{\delta_{\text{ZN}}}{\pi}\ln\left(\frac{\delta_{\text{ZN}}}{\pi}\right) - \arg\Gamma\left(i\frac{\delta_{\text{ZN}}}{\pi}\right) - \frac{\pi}{4}, \tag{4.40}$$

$$\delta_{\text{ZN}} = \frac{\pi}{16ab}\frac{\sqrt{6 + 10\sqrt{1 - b^{-4}}}}{1 + \sqrt{1 - b^{-4}}}. \tag{4.41}$$

The correction factor h_1 is introduced empirically [6]. The expression of δ_{ZN} is derived by using the linear potential model [8]. It should be noted that P_{12} becomes zero, when $\psi_{\text{ZN}} \equiv \sigma_{\text{ZN}} - \bar{\phi}_{\text{S}} = (n+1/2)\pi$ ($n = 0, 1, 2, \ldots$) holds. Namely, the intriguing phenomenon of complete reflection occurs, when this condition is satisfied.

When the phases are needed, the transition matrix I_X at $E \geq E_{\rm b}$ defined below should be used,

$$I_X = \begin{pmatrix} \sqrt{1-p_{\rm ZN}}e^{i\phi_{\rm S}} & \sqrt{p_{\rm ZN}}e^{i\sigma_0^{\rm ZN}} \\ -\sqrt{p_{\rm ZN}}e^{-i\sigma_0^{\rm ZN}} & \sqrt{1-p_{\rm ZN}}e^{-i\phi_{\rm S}} \end{pmatrix}, \tag{4.42}$$

where $\phi_{\rm S}$ is the same as Eq. (4.40). At $E > E_2(\infty)$, this I_X is used in the same way as I_X in the Landau-Zener case. When the trapping by the upper adiabatic potential occurs ($E \leq E_2(\infty)$), we have to use the reduced scattering matrix ($S_{NT}^{{\rm R}(a)}$) given by Eq. (4.16) with Eq. (4.17) that describes the overall transmission/reflection amplitudes from the entrance to the exit. The factors Δ_{ij} in Eq. (4.16) are defined as follows:

$$\Delta_{12} = \sigma_{\rm ZN}, \tag{4.43}$$

$$\Delta_{11} = 2\int_{T_2^l}^{R_{\rm b}} K_2(R)dR - 2\sigma_0^{\rm ZN}, \tag{4.44}$$

$$\Delta_{22} = 2\int_{R_{\rm b}}^{T_2^r} K_2(R)dR + 2\sigma_0^{\rm ZN}, \tag{4.45}$$

$$\sigma_0^{\rm ZN} = \left(\frac{R_{\rm b}-R_{\rm t}}{2}\right)\left\{K_1(R_{\rm t}) + K_2(R_{\rm b}) + \frac{1}{3}\frac{[K_1(R_{\rm t})-K_2(R_{\rm b})]^2}{K_1(R_{\rm t})+K_2(R_{\rm b})}\right\}, \tag{4.46}$$

and $\sigma_{\rm ZN}$ [$\phi_{\rm S}$] is given by Eq. (4.37) [Eq. (4.40)]. The index of the $S_{NT}^{{\rm R}(a)}$-matrix corresponds to left or right side of the barrier.

4.3 Classically Forbidden Transition

4.3.1 *Landau-Zener Type* ($E \leq E_2(R_0)$)

The Stokes constant $U_1[\equiv {\rm Re}(U_1) + i{\rm Im}(U_1)]$ is given by

$${\rm Re}\, U = \cos(\sigma_{\rm ZN})\left\{\sqrt{B(\sigma_{\rm ZN}/\pi)}e^{\delta_{\rm ZN}} - g_2\sin^2(\sigma_{\rm ZN})\frac{e^{-\delta_{\rm ZN}}}{\sqrt{B(\sigma_{\rm ZN}/\pi)}}\right\} \tag{4.47}$$

and

$$\operatorname{Im} U = \sin(\sigma_{\mathrm{ZN}})\left\{ B(\sigma_{\mathrm{ZN}}/\pi)e^{2\delta_{\mathrm{ZN}}} - g_2^2 \sin^2(\sigma_{\mathrm{ZN}})\cos^2(\sigma_{\mathrm{ZN}})\frac{e^{-2\delta_{\mathrm{ZN}}}}{B(\sigma_{\mathrm{ZN}}/\pi)} + 2g_1\cos^2(\sigma_{\mathrm{ZN}}) - g_2 \right\}^{1/2}, \tag{4.48}$$

where

$$B(x) = \frac{2\pi x^{2x}e^{-2x}}{x\Gamma^2(x)}, \tag{4.49}$$

$$g_1 = \frac{3\sigma_{\mathrm{ZN}}}{\pi\delta_{\mathrm{ZN}}}\ln(1.2 + a^2) - \frac{1}{a^2} \tag{4.50}$$

and

$$g_2 = 1.8(a^2)^{0.23}e^{-\delta_{\mathrm{ZN}}}. \tag{4.51}$$

The parameters σ_{ZN} and δ_{ZN} are given below (Eqs. (4.60)–(4.63)). The correction factors g_1 and g_2 are originally unity. The correction g_1 is first determined by fitting the nonadiabatic transition probability p_{ZN} to the exact one in the linear model and then the correction g_2 is determined by using the total transition probability P_{12} [6]. The first factor of g_1 is responsible for the weak coupling case ($a^2 \gg 1$) and the second factor takes care of the strong coupling case ($a^2 \ll 1$) at $b^2 \sim 0^-$.

The overall transition probability from state 1 to state 2 is given by

$$P_{12} = 4p_{\mathrm{ZN}}(1 - p_{\mathrm{ZN}})\sin^2(\psi_{ZN}) \tag{4.52}$$

with

$$p_{\mathrm{ZN}} = \frac{1}{1 + |U_1|^2} = [1 + B(\sigma_{\mathrm{ZN}}/\pi)\exp(2\delta_{\mathrm{ZN}}) - g_1\sin^2(\sigma_{\mathrm{ZN}})]^{-1} \tag{4.53}$$

and

$$\psi_{\mathrm{ZN}} = \arg(U_1). \tag{4.54}$$

In the case of $b^2 \geq 0$ this equals $\phi_{\mathrm{S}} + \sigma_0^{\mathrm{ZN}}$ and the latter two factors are given by Eqs. (4.24) and (4.26).

When the dynamical phases induced by the nonadiabatic transition are needed, the reduced scattering matrix ($S_{LZ}^{\mathrm{R}(a)}$) given by Eq. (4.14), which provides the transition amplitude from the turning point on the entrance adiabatic surface to that on the exit adiabatic surface, should be used:

$$S_{LZ}^{\mathrm{R(a)}} = I^{\mathrm{T}} \cdot I, \tag{4.55}$$

where the superscript T means "transpose" and the transition matrix I is given by

$$I = \begin{pmatrix} \sqrt{1 - p_{ZN}} \exp[i(\psi_{ZN} - \sigma_{ZN})] & -\sqrt{p_{ZN}} \exp(i\sigma_{ZN}) \\ \sqrt{p_{ZN}} \exp(-i\sigma_{ZN}) & \sqrt{1 - p_{ZN}} \exp[-i(\psi_{ZN} - \sigma_{ZN})] \end{pmatrix}. \tag{4.56}$$

It should be noted that this matrix I is different from the matrix I_X which represents the nonadiabatic transition at the crossing point and is given by Eq. (4.22). It should be noted that the elements of the reduced scattering matrix $S_{LZ}^{\mathrm{R}(a)}$ are expressed in a similar way as in the classically allowed case. They are explicitly given by

$$\left(S_{LZ}^{\mathrm{R}(a)}\right)_{11} = \left[p_{ZN} + (1 - p_{ZN})e^{2i\psi_{ZN}}\right]e^{-2i\sigma_{ZN}}, \tag{4.57}$$

$$\left(S_{LZ}^{\mathrm{R}(a)}\right)_{22} = \left[p_{ZN} + (1 - p_{ZN})e^{-2i\psi_{ZN}}\right]e^{2i\sigma_{ZN}} \tag{4.58}$$

and

$$\left(S_{LZ}^{\mathrm{R}(a)}\right)_{12} = \left(S_{LZ}^{\mathrm{R}(a)}\right)_{21} = -2i\sqrt{p_{ZN}(1 - p_{ZN})}\sin\psi_{ZN}. \tag{4.59}$$

The parameters σ_{ZN} and δ_{ZN} are given as follows:

(1) At energies in between $E_1(R_0)$ and $E_2(R_0)$ ($E_2(R_0) > E > E_1(R_0)$),

$$\sigma_{\mathrm{ZN}} = \int_{T_1}^{R_0} K_1(R)dR + \sigma_0^{\mathrm{ZN}} \tag{4.60}$$

and

$$\delta_{\mathrm{ZN}} = \int_{R_0}^{T_2} K_2(R)dR + \delta_0^{\mathrm{ZN}}. \tag{4.61}$$

(2) At energies lower than $E_1(R_0)$ $(E \leq E_1(R_0))$,

$$\sigma_{\mathrm{ZN}} = \sigma_0^{\mathrm{ZN}} \tag{4.62}$$

and

$$\delta_{\mathrm{ZN}} = -\int_{R_0}^{T_1} |K_1(R)|dR + \int_{R_0}^{T_2} |K_2(R)|dR + \delta_0^{\mathrm{ZN}}. \tag{4.63}$$

4.3.2 *Nonadiabatic Tunneling Type*

4.3.2.1 At Energies in between E_t and E_b $(E_\mathrm{b} \geq E \geq E_\mathrm{t})$

The Stokes constant U_1 is given by

$$U_1 = i\frac{\sqrt{1+W^2}e^{i\phi} - 1}{W} \tag{4.64}$$

with

$$W = \frac{h_2}{a^{2/3}} \int_0^\infty \cos\left[\frac{t^3}{3} - \frac{b^2}{a^{2/3}}t - \frac{h_3}{a^{2/3}}\frac{t}{h_4 + a^{1/3}t}\right] dt \tag{4.65}$$

and

$$\phi = \sigma_{\mathrm{ZN}} + \arg\Gamma\left(\frac{1}{2} + i\frac{\delta_{\mathrm{ZN}}}{\pi}\right) - \frac{\delta_{\mathrm{ZN}}}{\pi}\ln\left(\frac{\delta_{\mathrm{ZN}}}{\pi}\right) + \frac{\delta_{\mathrm{ZN}}}{\pi} - h_5, \tag{4.66}$$

where

$$h_2 = 1 + \frac{0.38}{a^2}(1+b^2)^{1.2-0.4b^2}, \tag{4.67}$$

$$h_3 = \frac{\sqrt{a^2 - 3b^2}}{\sqrt{a^2+3}}\sqrt{1.23 + b^2}, \tag{4.68}$$

$$h_4 = 0.61\sqrt{2+b^2}, \tag{4.69}$$

$$h_5 = 0.34\frac{a^{0.7}(a^{0.7}+0.35)}{a^{2.1}+0.73}(0.42+b^2)\left(2 + \frac{100b^2}{100+a^2}\right)^{\frac{1}{4}}, \tag{4.70}$$

$$\sigma_{\mathrm{ZN}} = -\frac{1}{\sqrt{a^2}}\left[0.057(1+b^2)^{\frac{1}{4}} + \frac{1}{3}\right](1-b^2)\sqrt{5+3b^2} \tag{4.71}$$

and

$$\delta_{\mathrm{ZN}} = \frac{1}{\sqrt{a^2}}\left[0.057(1-b^2)^{\frac{1}{4}} + \frac{1}{3}\right](1+b^2)\sqrt{5-3b^2}. \tag{4.72}$$

The correction factors h_2, h_3 and h_4 are originally 1/2, 1, and 1, respectively, and are confirmed to cover the whole range of a^2 [6]. The corrections used for h_5, $\sigma_{\rm ZN}$ and $\delta_{\rm ZN}$ are introduced by using the linear potential model [6].

The physically most meaningful quantity is the overall transmission probability (P_{12}) from left to right or vice versa and is given by

$$P_{12} = \frac{W^2}{1+W^2}. \tag{4.73}$$

When the phases are needed, what we have to use is the reduced scattering matrix ($S_{NT}^{\mathrm{R}(a)}$) defined by Eq. (4.16) with Eq. (4.17). The parameters $\Delta_{12}, \Delta_{11}, \Delta_{22}$ are defined as follows:

$$\Delta_{12} = \sigma_{\rm ZN}, \qquad \Delta_{11} = \sigma_{\rm ZN} - 2\sigma_0^{\rm ZN} \quad \text{and} \quad \Delta_{22} = \sigma_{\rm ZN} + 2\sigma_0^{\rm ZN}, \tag{4.74}$$

where

$$\sigma_0^{\rm ZN} = -\frac{1}{3}(R_{\rm t} - R_{\rm b})K_1(R_{\rm t})(1+b^2). \tag{4.75}$$

4.3.2.2 At Energies Lower than E_t ($E \leq E_t$)

The Stokes constant is given by

$$\mathrm{Re}\, U_1 = \sin(2\sigma_c)\left\{\frac{0.5\sqrt{a^2}}{1+\sqrt{a^2}}\sqrt{B\left(\frac{\sigma_c}{\pi}\right)}e^{-\delta_{\rm ZN}} + \frac{e^{\delta_{\rm ZN}}}{\sqrt{B(\sigma_c/\pi)}}\right\} \tag{4.76}$$

and

$$\mathrm{Im}\, U_1 = \cos(2\sigma_c)\sqrt{\frac{(\mathrm{Re}\, U_1)^2}{\sin^2(2\sigma_c)} + \frac{1}{\cos^2(2\sigma_c)}} - \frac{1}{2\sin(\sigma_c)}\left|\frac{\mathrm{Re}\, U_1}{\cos(\sigma_c)}\right|, \tag{4.77}$$

where

$$\sigma_c = \sigma_{\rm ZN}(1 - 0.32 \times 10^{-2/a^2} e^{-\delta_{\rm ZN}}), \tag{4.78}$$

$$\delta_{\rm ZN} = \int_{T_1^{\rm l}}^{T_1^{\rm r}} |K_1(R)|dR \tag{4.79}$$

and

$$\sigma_{\mathrm{ZN}} = \frac{\pi}{8a|b|}\frac{1}{2}\frac{\sqrt{6+10\sqrt{1-1/b^4}}}{1+\sqrt{1-1/b^4}}. \tag{4.80}$$

The function $B(x)$ is defined by Eq. (4.49).

The physically meaningful overall transmission probability P_{12} is given by

$$P_{12} = \frac{B(\sigma_c/\pi)e^{-2\delta_{\mathrm{ZN}}}}{[1+(0.5\sqrt{a^2}/[\sqrt{a^2}+1])B(\sigma_c/\pi)e^{-2\delta_{\mathrm{ZN}}}]^2 + B(\sigma_c/\pi)e^{-2\delta_{\mathrm{ZN}}}}. \tag{4.81}$$

It should be noted that when $a^2 \to 0$, namely, the upper adiabatic potential goes away, we have

$$P_{12} = \frac{e^{-2\delta_{\mathrm{ZN}}}}{1+e^{-2\delta_{\mathrm{ZN}}}}, \tag{4.82}$$

which agrees with the ordinary single potential barrier penetration probability. The factor $0.5\sqrt{a^2}/(1+\sqrt{a^2})$ in Eq. (4.81) is originally 0.25 and later confirmed to be better [6]. The factor $0.32 \times 10^{-2/a^2}$ in Eq. (4.78) is originally 0 and later confirmed to be better [6]. The simplified expression of σ_{ZN} is derived by using the linear potential model [8].

When the phases are needed, we have to use the reduced scattering matrix $S_{NT}^{\mathrm{R}(a)}$ in the same way as above and the phases Δ_{ij} are given by

$$\Delta_{12} = \Delta_{11} = \Delta_{22} = -2\sigma_{\mathrm{ZN}}. \tag{4.83}$$

4.4 How to Use the Theory in Multi-Dimensional Dynamics

The Zhu-Nakamura formulas presented above are valid for one-dimensional systems. In the case of multi-dimensional dynamics some care is needed naturally. The procedure is briefly explained here. The details can be found in [1–3].

4.4.1 Classically Allowed Case

Along a classical trajectory satisfying the initial condition of quantum state with the total energy E, the energy difference ΔE between the current adiabatic potential energy surface and the neighboring adiabatic potential energy surface is monitored and the position of minimum ΔE is detected. Calculate the nonadiabatic coupling vector there. Cut the two adiabatic potential energy surfaces along this vector to prepare the two adiabatic potential energy curves $E_j(R)$ $(j = 1, 2)$ along this direction, where R is the coordinate along this direction. Then determine the transition type, i.e., Landau-Zener type or nonadiabatic tunneling type, and compute the hopping probability p_{ZN} by using the appropriate formula of the Zhu-Nakamura theory. This probability is nothing but the nonadiabatic transition probability for one passage of the avoided crossing point [see Eqs. (4.19) and (4.35)]. In the case of ant-eater procedure, the trajectory hops to the other potential surface when p_{ZN} is larger than the random number p_{random} $(0 \leq p_{\mathrm{random}} \leq 1)$ generated there. Otherwise, the trajectory stays on the same potential energy surface.

After the transition the new momentum is given by

$$\frac{1}{2m}p_{\mathrm{aft}}^2 = \frac{1}{2m}p_{\mathrm{bfr}}^2 - [E_2(R_0) - E_1(R_0)], \tag{4.84}$$

where p_{bfr} is the momentum before the transition. The momenta p_{bfr} and p_{aft} are the components along the direction of transition. The other components after the transition are the same as those before the transition.

The nonadiabatic coupling vector needed to determine the direction of transition should be estimated quantum chemically. If it is too CPU-time consuming, however, the following approximation can be employed [2, 9]. In the vicinity of conical intersection we can assume the following diabatic Hamiltonian,

$$H = \begin{pmatrix} V_1 & V \\ V & V_2 \end{pmatrix} = \begin{pmatrix} \sum_j A_j x_j & \sum_j B_j x_j \\ \sum_j B_j x_j & -\sum_j A_j x_j \end{pmatrix}. \tag{4.85}$$

Then the nonadiabatic coupling vector $\vec{T} \equiv (T_1, T_2, \cdots)$ is given by[1]

$$T_j \simeq \frac{\frac{\partial V}{\partial x_j}(V_1 - V_2) - V\left(\frac{\partial V_1}{\partial x_j} - \frac{\partial V_2}{\partial x_j}\right)}{(\Delta E)^2}$$

$$= 2\frac{\sum_k (A_k B_j - A_j B_k)x_k}{(\Delta E)^2} \equiv \frac{2}{(\Delta E)^2} e_j, \tag{4.86}$$

where

$$\Delta E = 2\sqrt{\left(\sum_k A_k x_k\right)^2 + \left(\sum_k B_k x_k\right)^2}. \tag{4.87}$$

On the other hand, the second derivative of ΔE is given by

$$\frac{\partial^2 \Delta E}{\partial x_i \partial x_j} = \frac{4}{(\Delta E)^3}\left[\left(\sum_k A_k x_k\right)^2 B_i B_j + \left(\sum_k B_k x_k\right)^2 A_i A_j \right.$$

$$\left. - \left(\sum_k A_k x_k\right)\left(\sum_k B_k x_k\right)(A_i B_j + B_i A_j)\right]. \tag{4.88}$$

Thus, up to an irrelevant scale factor we have

$$\frac{\partial^2 \Delta E}{\partial x_i \partial x_j} \propto e_i e_j. \tag{4.89}$$

This gives a rank-1 matrix, since all columns are linearly dependent; thus, the eigenvector of its only non-zero eigenvalue provides the direction of the nonadiabatic coupling vector. This approximation was demonstrated to work well [2, 9].

4.4.2 *Classically Forbidden Case*

In this case the avoided crossing point is located in the energetically inaccessible region, thus the quantum mechanical tunneling through an adiabatic potential barrier is required on the way in to reach the transition point as well as on the way out from the transition point

[1] In the first one of [2] there are typographical errors in Eqs. (10.8)–(10.10). They should read as described here.

after the transition. These two effects, i.e., nonadiabatic transition and quantum mechanical tunneling, cannot be separated. Besides, the optimal tunneling path is not a straight line in general, although the nonadiabatic transition itself is very much localized at the avoided crossing point. In this sense, the Zhu-Nakamura formulas along the straight line in the direction of the nonadiabatic coupling vector at the caustic C cannot be accurate in multi-dimensional space. Fortunately, however, the parameters σ_{ZN} and δ_{ZN} in the Zhu-Nakamura formulas of P_{ZN} given by Eqs. (4.73) and (4.81) clearly represent the effects of nonadiabatic transition and quantum mechanical tunneling, respectively. Thus, these effects can be evaluated separately. This means that we can find the optimal paths in the classically forbidden regions by *maximizing* the overall transition probabilities P_{ZN} given by the Zhu-Nakamura theory. The method to determine the tunneling path is basically the same as that explained in [10] (see also [3]). As the zeroth-order approximation of the tunneling path to start with, the straight line in the steepest ascent direction or the direction of nonadiabatic coupling vector at the caustic is employed. The schematic view is shown in Fig. 4.3. The blue line from C(caustic) to Q_0 through P_0 is the zeroth-order tunneling path and the purple line from C to Q through P is the optimal path that maximizes the overall nonadiabatic tunneling probability. The orange circle schematically represents the nonadiabatic transition on the way.

The nonadiabatic tunneling type and the Landau-Zener type are discussed separately in the following subsections.

4.4.2.1 Nonadiabatic tunneling type

The case of $E(= E_l) < E_t$ is considered first (see Fig. 4.4). The other classically forbidden case $E_b > E(= E_m) > E_t$ is explained at the end of this subsection.

Starting from the caustic, the tunneling path runs through the potential barrier of the lower adiabatic potential $E_1(\mathbf{q})$. The imaginary part of the action consists of two parts. Its mth-order approximation is expressed as

$$S^{(m)} = S_m^{\mathrm{CP}} + S_m^{\mathrm{PQ}}, \tag{4.90}$$

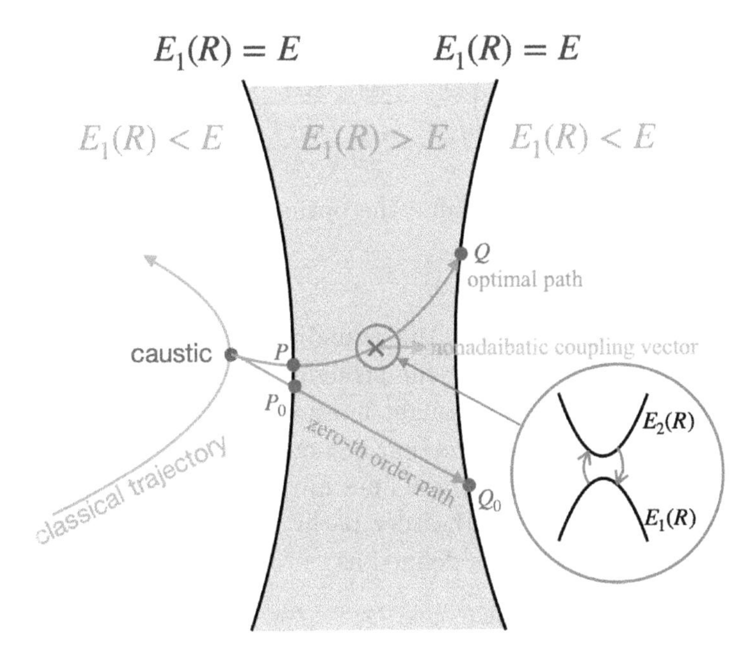

Figure 4.3: Schematic view of the zeroth-order and the optimal nonadiabatic tunneling path. Reproduced with permission from [3].

where P(Q) is the crossing point of the path with the equi-potential surface at the entrance to (exit from) the tunneling region. The coordinates $\{q_j^{(m)}\}$ of the tunneling path are expressed as

$$q_j^{(m)}(z) = q_j^{\mathrm{C}} + \sum_{k=1}^{N_I} C_{jk}^{(m)} z^k, \tag{4.91}$$

where $\{q_j^{\mathrm{C}}\}$ represent the coordinates at the caustic C, $C_{jk}^{(m)}$ are the expansion coefficients to be determined and the variable z runs from $z = 0$ at the caustic C to $z = 1$ at the tunnel exit Q. From energy conservation,

$$E = \frac{1}{2}\sum_{j=1}^{N_I}(\dot{q}_j^{(m)})^2 + E_1(\mathbf{q}) = \frac{1}{2}\sum_{j=1}^{N_I}\left(\frac{dq_j^{(m)}}{dz}\right)^2 (\dot{z})^2 + E_1(\mathbf{q}), \tag{4.92}$$

we have

$$(\dot{z})^2 \equiv \left(\frac{dz}{dt}\right)^2 = -\frac{2(E_1(\mathbf{q}) - E)}{\sum_j \left(\frac{dq_j^{(m)}}{dz}\right)^2} \equiv -\zeta^2. \tag{4.93}$$

The procedure to determine the optimal path is summarized as follows.

(i) Step 1

When the trajectory enters the nonadiabatic transition zone, the first caustic is detected and the straight line is drawn from there. As mentioned above, two straight lines are used: in the direction of steepest ascent and the direction of nonadiabatic coupling vector. The action integrals along these lines are calculated and the smaller one is selected as the zeroth-order path. The unit vector $\vec{e}_{\mathrm{SL}}$ along the selected straight line is defined as

$$\vec{e}_{\mathrm{SL}} = (a_1, a_2, \cdots, a_{N_I}), \tag{4.94}$$

where $\{a_k\}$ are direction cosines and in the case of steepest ascent these are given by

$$a_k = \frac{\partial E_1/\partial q_k}{\sqrt{\sum_k (\partial E_1/\partial q_k)^2}}. \tag{4.95}$$

Here $E_1 = E_1(\mathbf{q})$ is the lower adiabatic potential. The new axis which starts from the caustic and runs along the straight line is denoted as $\tilde{q}_1$ whose components measured in the original space $\{q_j\}$ are given as

$$q_k^{\mathrm{SL}} \equiv (\tilde{q}_1)_k = q_k^{\mathrm{C}} + a_k\xi, \tag{4.96}$$

where ξ is a parameter with $\xi = 0$ corresponding to the caustic. This straight line is extended and the crossing point P_0 with the equi-potential surface, i.e., the entrance to the tunneling region, should be detected. The zeroth-order action integral along this line is evaluated as

$$S_0^{\mathrm{SL}}(\xi) \equiv S_0^{\mathrm{CP_0}} + S_0^{\mathrm{P_0}}(\xi), \tag{4.97}$$

where

$$S_0^{\mathrm{CP_0}} = \sum_{k=0}^{N_I} \int_{\mathrm{C}}^{\mathrm{P_0}} p_k^{\mathrm{SL}} dq_k^{\mathrm{SL}} = \sum_{k=0}^{N_I} (a_k)^2 \int_0^{\xi_{\mathrm{P}}} \tilde{\mathrm{p}}_1 d\xi \tag{4.98}$$

and

$$S_0^{\mathrm{P}_0}(\xi) = \sum_{k=0}^{N_I} \int_{\mathrm{P}_0}^{\xi} p_k^{\mathrm{SL}} dq_k^{\mathrm{SL}} = \sum_{k=0}^{N_I} (a_k)^2 \int_{\xi_{\mathrm{P}_0}}^{\xi} \tilde{p}_1 d\xi. \tag{4.99}$$

Here $\xi_{\mathrm{P}_0} = \xi(\mathrm{P}_0)$, $p_k^{\mathrm{SL}} = a_k \tilde{p}_1$ and $\tilde{p}_1$ is the momentum along this straight line defined as

$$\tilde{p}_1 = \begin{cases} \sqrt{2}\sqrt{V_{\mathrm{I}} - V_{\mathrm{C}}} & (\text{for } 0 \le \xi \le \xi_{\mathrm{P}_0}) \\ \sqrt{2}\sqrt{V_{\mathrm{I}} - E} & (\text{for } \xi_{\mathrm{P}_0} \le \xi), \end{cases} \tag{4.100}$$

where $V_{\mathrm{C}} = E_1(R_{\mathrm{C}})$ is the potential at C and V_{I} is the potential along $\tilde{q}_1$,

$$V_{\mathrm{I}} = E_1\big(\tilde{q}_1(\xi), \{\tilde{q}_j\}_{j \neq 1}\big). \tag{4.101}$$

The action $S_0^{\mathrm{P}_0}(\xi)$ from the point P_0 is evaluated step by step. When the total action S_0^{SL} becomes bigger than a certain criterion,

$$S_0^{\mathrm{SL}}(\xi) \ge Bigact, \tag{4.102}$$

then the tunneling there is not carried out and the classical trajectory is further propagated. If the straight line reaches the equi-potential surface in the exit channel at $\xi = \xi_{\mathrm{Q}_0}$ and the total action is smaller than the above criterion, then the determination of the optimal tunneling path is carried out.

The effect of the nonadiabatic transition on the way is taken into account as follows: On the straight line the minimum adiabatic energy difference position $q_{\mathrm{tr}}^{(0)}(z = z_0)$ is detected and the normalized nonadiabatic coupling vector $\mathbf{e}_{\mathrm{nad}}$ is evaluated there.[2] Along the direction of this vector two adiabatic potential curves $E_1(R)$ and $E_2(R)$ are calculated and the parameters a^2 and b^2 are evaluated from Eqs. (4.8) and (4.9). The total energy E in the expression of b^2 is replaced by the effective total energy along the direction of the

[2] Strictly speaking, q_{tr} should be equal to $\mathrm{Re}(R_*)$, where R_* is the complex crossing point, but the position of the minimum adiabatic energy difference is good enough for practical computations.

nonadiabatic coupling vector E_{eff}, which is given by

$$E_{\text{eff}} = -\left[\frac{1}{2}\left(\sum_j \frac{dq_j^{(m=0)}}{dz} e_j\right)^2 (\dot{z})^2 + E_1(\mathbf{q})\right]_{z=z_0}, \tag{4.103}$$

where e_j is the jth component of the coupling vector $\mathbf{e}_{\text{nad}}$. The overall transition probability denoted as $P_{\text{ZN}}^{(0)}$ is calculated from Eq. (4.81), where $\delta_{\text{ZN}} = S^{(0)} \equiv S_0^{\text{SA}}(\xi_{\text{Q}_0})$ and σ_{ZN} is given by Eq. (4.80). If this probability is smaller than a certain critical value, say ϵ,

$$P_{\text{ZN}}^{(0)} \leq \epsilon, \tag{4.104}$$

then the search of the optimal path is stopped here and the classical trajectory is further propagated to detect the next caustic. If $P_{\text{ZN}}^{(0)}$ is larger than the criterion ϵ, then we proceed to find the optimal path. The first-order path is provided by

$$C_{j1}^{(1)} = C_{j1}^{(0)}, \; C_{jn}^{(1)} = \text{small number} \quad \text{for } n \geq 2. \tag{4.105}$$

Then the first-order action integral $S^{(m=1)}$ is calculated. The method of calculation of $S^{(m)}$ is described below.

(ii) Step 2

The tunneling path $\{q_j^{(m)}\}$ and the corresponding probability $P_{\text{ZN}}^{(m)}$ in the mth-order approximation are calculated as explained below. The expansion coefficients $C_{jk}^{(m)}$ for $m \geq 2$ are obtained by slightly modifying the coefficients in the $(m-1)$-th-order approximation as

$$C_{jk}^{(m)} = C_{jk}^{(m-1)} + \text{sgn} \cdot \Delta_{jk}^{(m)}, \tag{4.106}$$

where $\Delta_{jk}^{(m)}$ is a positive small number and

$$\text{sgn} = +1 \text{ when } \frac{\partial P_{\text{ZN}}^{(m-1)}}{\partial C_{jk}^{(m-1)}} > 0, \quad -1 \text{ when } \frac{\partial P_{\text{ZN}}^{(m-1)}}{\partial C_{jk}^{(m-1)}} < 0. \tag{4.107}$$

Since the end point of the path at $z = 1$ is not necessarily located on the equi-potential surface $[E = E_1(z) \equiv E_1(\mathbf{q}^{(m)}(z = 1))]$, we introduce a scaling factor α to correct this,

$$z \to \alpha z, \tag{4.108}$$

where

$$\alpha <, =, > 1 \text{ for } E_1(z = 1) <, =, > E. \tag{4.109}$$

Then the new coefficients are modified as

$$C_{jk}^{(m)} \to C_{jk}^{(m)} \times \alpha^n. \tag{4.110}$$

The actions S_m^{CP} and S_m^{PQ} are calculated separately (see Eq. (4.90)). In the case of S_m^{CP} the separability between the direction along the straight line and the direction normal to that is assumed. Then, the pure imaginary action S_m^{CP} is obtained from

$$S_m^{\mathrm{CP}} = -2i \int K_\perp dt, \tag{4.111}$$

where $K_\perp$ is the kinetic energy in the direction of $\tilde{q}_1$ given as

$$K_\perp = \frac{1}{2}\sum_{j=1}^{N_I} (\dot{q}_{j\perp})^2 = \frac{1}{2}\sum_j (q_j')^2 (a_j)^2 (\dot{z})^2 \equiv K_\perp^{(0)} (\dot{z})^2, \tag{4.112}$$

where

$$\dot{q}_j(z) = q_j' \dot{z} \equiv \frac{dq_j}{dz}\dot{z}. \tag{4.113}$$

From energy conservation (see Eq. (4.92))

$$V_C = K_\perp + E_1(\mathbf{q}^{\mathrm{SA}}) = K_\perp^{(0)}(\dot{z})^2 + E_1(\mathbf{q}^{\mathrm{SA}}), \tag{4.114}$$

we have

$$\dot{z} = i\sqrt{\frac{E_1(\mathbf{q}^{\mathrm{SA}}) - V_C}{K_\perp^{(0)}}} \equiv i\zeta(z). \tag{4.115}$$

Then the action S_m^{CP} is finally given by

$$S_m^{\mathrm{CP}} = -2i\int K_\perp^{(0)}\dot{z}dz = 2\int K_\perp^{(0)}\zeta(z)dz$$
$$= 2\int K_\perp^{(0)}\sqrt{\frac{E_1(\mathbf{q}^{\mathrm{SA}}) - V_C}{K_\perp^{(0)}}}dz$$
$$= \int_0^{z_P}\left[2\sum_j \left(q_j'(z)a_j\right)^2\right]^{1/2}\sqrt{E_1(\mathbf{q}^{\mathrm{SA}}) - V_C}dz, \quad (4.116)$$

where z_{P} is the value of z at the equi-potential position P.

The action S_m^{PQ} is evaluated as follows: In the region in between P and Q all directions are classically forbidden. The energy conservation and the parameter $\dot{z}$ are given by Eqs. (4.92) and (4.93). By using the same method as above, the action S_m^{PQ} is obtained as

$$S_m^{\mathrm{PQ}} = -2i\int Kdt = \int_{z_{\mathrm{P}}}^{z=1}\left[2(E_1(z) - E)\left(\sum_j (q_j')^2\right)\right]^{1/2} dz. \quad (4.117)$$

Next, we have to consider the effect of nonadiabatic transition. The minimum energy separation position $q_{\mathrm{tr}}^{(m)}$ and the basic parameters ($a^2, b^2, \sigma_{\mathrm{ZN}}$ and δ_{ZN}) are evaluated in the same way as above. Then, finally the overall transmission probability $P_{\mathrm{ZN}}^{(m)}$ is calculated from Eq. (4.81), where the effective total energy E_{eff} at $q_{\mathrm{tr}}^{(m)}$ is given by Eq. (4.103) for $m \neq 0$.

(iii) Step 3

The above procedure to find the optimal path is repeated until the required convergence of the probability P_{ZN} is attained. If it is converged, a random number p_{random} is generated and if

$$P_{\mathrm{ZN}} \geq p_{\mathrm{random}} \quad (4.118)$$

is satisfied, then the transition, i.e., the "nonadiabatic transmission" is decided to occur and a new trajectory starts from Q with zero initial momenta. If the condition Eq. (4.118) is not satisfied, the original trajectory before the transition is further propagated.

If the energy $E(= E_m)$ is in between the top E_t of the lower adiabatic potential and the bottom E_b of the upper adiabatic potential (see Fig. 4.4), the transition itself is classically forbidden, but no actual tunneling through the lower adiabatic potential $E_1(\mathbf{q})$ is involved and the nonadiabatic transmission probability $P_{\rm ZN}$ given by Eq. (4.73) can be used instead of $p_{\rm ZN}$ in the procedure of

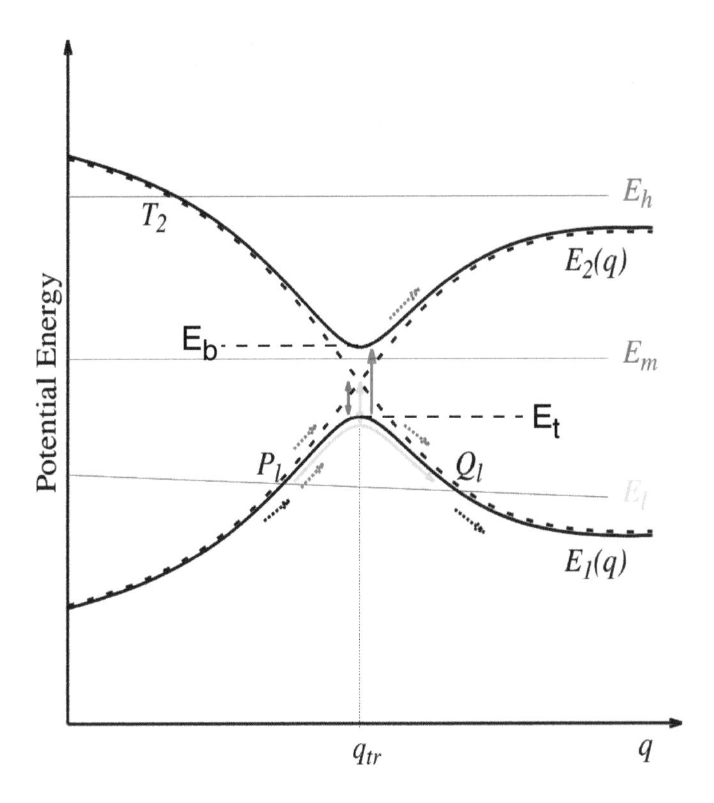

Figure 4.4: Schematic view of the hopping algorithm in the case of nonadiabatic tunneling type. Solid (dotted) arrows indicate the nonadiabatic transitions (adiabatic propagations). Case (1) (red arrow): vertical transition at $q_{\rm tr}$ at $E = E_h$. Case (2) (blue arrow): non-classical transition at $q_{\rm tr}$ at $E = E_m$. Case (3) (green arrow): non-classical horizontal transition from P_l to Q_l at $E = E_l$. The optimal tunneling path from P_l to $q_{\rm tr}$ and $q_{\rm tr}$ to Q_l is determined variationally in the multi-dimensional space.

the classically allowed case. The new trajectory starts from $q_{\rm tr}$ with the initial momentum along this transition direction given by $\sqrt{2m(E_m - E_t)}$. The momentum components in other directions are the same as those before the transition.

4.4.2.2 Landau-Zener type

The situation in this case is a bit more complicated than in the case of nonadiabatic tunneling type, since the tunneling path jumps from $E_1(\mathbf{q})$ to $E_2(\mathbf{q})$ at the minimum adiabatic energy difference position $q_{\rm tr}$. In the semiclassical Zhu-Nakamura theory of nonadiabatic transition, the transition is assumed to occur locally at the avoided crossing point and thus the processes before and after the transition can be treated separately. This means that the second portion of the whole tunneling path, namely the path from $q_{\rm tr}$ to the tunnel exit Q on the adiabatic potential $E_2(\mathbf{q})$, can be determined separately from the first one in each iteration process.

Let us first consider the case $E(= E_l) < E_1(q_{\rm tr})$ (see Fig. 4.5). The overall transition is depicted by the green line from P_l to Q_l, where three events are involved: tunneling from P_l to $q_{\rm tr}$ on $E_1(\mathbf{q})$, nonadiabatic transition at $q_{\rm tr}$ from $E_1(q_{\rm tr})$ to $E_2(q_{\rm tr})$, and tunneling from $q_{\rm tr}$ to Q_l on $E_2(q_{\rm tr})$. In the zeroth-order approximation, the straight line from the caustic C ($z_1 = 0$) on $E_1(\mathbf{q})$ is extended to $q_{\rm tr}^{(0)}$ ($z_1 = 1$) and the second one is a straight line in the steepest *descent* direction from $q_{\rm tr}^{(0)}$ ($z_2 = 0$) on $E_2(\mathbf{q})$ to the exit Q_0 ($z_2 = 1$). In the general mth-order approximation, the second path from $q_{\rm tr}$ to the exit Q_l is determined so that the total transition probability $P_{\rm ZN}$ is maximized. Here the suffix m is omitted for simplicity. The first path from C to $q_{\rm tr}$ gives the action,

$$S_1 = S_1^{\rm CP} + S_1^{\rm Pq_{tr}} \tag{4.119}$$

and the second path from $q_{\rm tr}$ to Q_l provides the action,

$$S_2 = S_2^{\rm q_{tr}Q_{\mathit{l}}}. \tag{4.120}$$

The various quantities such as a^2, b^2, $\sigma_{\rm ZN}$ and $\delta_{\rm ZN}$ can be estimated from the expressions in Section 4.3.1. The effective total energy $E_{\rm eff}$ (see Eq. (4.103)) replaces E in Eq. (4.4). Then the parameters $\sigma_0^{\rm ZN}$

and δ_0^{ZN} can be calculated from Eq. (4.26). The phase integral δ_{ZN} is given by (see Eq. (4.63))

$$\delta_{\mathrm{ZN}} = -S_1 + S_2 + \delta_0^{\mathrm{ZN}}. \tag{4.121}$$

The overall transition probability P_{ZN} is given by Eq. (4.52). This process is repeated until the probability P_{ZN} converges. Other procedures are basically the same as explained in the nonadiabatic tunneling case.

In the case $E_1(q_{\mathrm{tr}}) < E(= E_m) < E_2(q_{\mathrm{tr}})$ (see Fig. 4.5), the overall transition is depicted by the blue line from q_{tr} to Q_m which is composed of the nonadiabatic transition at q_{tr} and the tunneling from q_{tr} to Q_m. The first part of the tunneling needed above is not necessary here. The parameters σ_{ZN} and δ_{ZN} should be evaluated

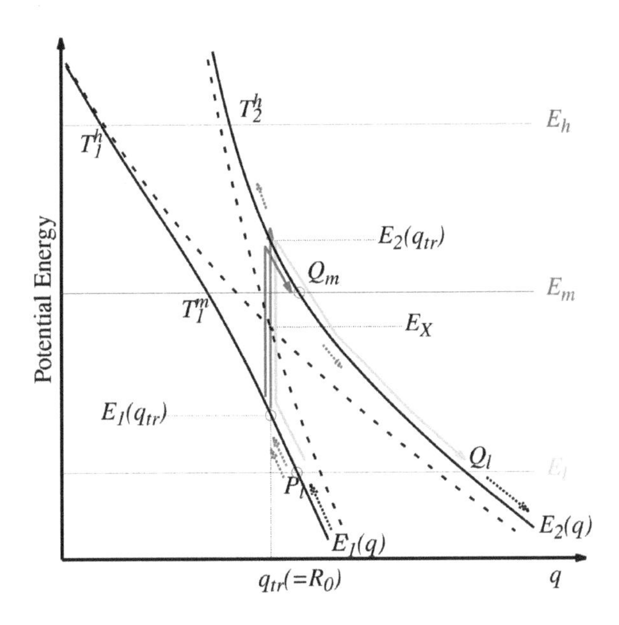

Figure 4.5: Schematic view of the hopping algorithm from $E_1(\mathbf{q})$ to $E_2(\mathbf{q})$ in the Landau-Zener type. Solid (dotted) arrows indicate the nonadiabatic transitions (adiabatic propagations). Case (1) (red arrow): vertical transition at q_{tr} at $E = E_h$. Case (2) (blue arrow): non-classical transition from q_{tr} to Q_m at $E = E_m$. Case (3) (green arrow): non-classical horizontal transition from P_l to Q_l at $E = E_l$. The optimal tunneling path from P_l to q_{tr} and q_{tr} to Q_l is determined variationally in the multi-dimensional space.

from Eqs. (4.60) and (4.61). Otherwise, the procedures are the same as above.

4.5 Time-Dependent Version of Formulas

The formulas derived in the time-independent framework can be easily transferred into the time-dependent version. It should be noted that the nonadiabatic tunneling type of transition does not show up and only the Landau-Zener type appears in the time-dependent problems, since time is unidirectional. The classically forbidden transitions in the time-independent framework correspond to the diabatically avoided crossing case in the time-dependent framework.

The following replacements of the parameters in the time-independent Zhu-Nakamura formulas are good enough for the transfer:

$$a^2 \Leftrightarrow \alpha = \frac{\sqrt{d^2-1}\hbar^2}{2V_0^2(t_t^2 - t_b^2)} \tag{4.122}$$

$$b^2 \Leftrightarrow \beta = -\sqrt{d^2-1}\frac{t_b^2 + t_t^2}{t_t^2 - t_b^2} \tag{4.123}$$

and

$$\sigma_{ZN} + i\delta_{ZN} = \frac{1}{\hbar}\left[\int_0^{t_b} E_+(t)dt - \int_0^{t_t} E_-(t)dt + \sqrt{\frac{\beta}{\alpha}} + \Delta_1\right] \tag{4.124}$$

with

$$\Delta_1 = \frac{t_0 - (t_b + t_t)/2}{\sqrt{\alpha(\beta^2 + i)}(t_b - t_t)}\sqrt{\frac{d^2}{d^2-1}} + \frac{1}{2\sqrt{\alpha}}\int_0^i \left(\frac{1+t^2}{t+\beta}\right)^{1/2} dt \tag{4.125}$$

$$V_0 = \frac{1}{2}(E_+(t_0) - E_-(t_0)) \tag{4.126}$$

and

$$d^2 = \frac{[E_+(t_b) - E_-(t_b)][E_+(t_t) - E_-(t_t)]}{[E_+(t_0) - E_-(t_0)]^2}, \tag{4.127}$$

where $E_\pm(t)$ are the adiabatic potentials with $E_+(t) > E_-(t)$, $t_b(t_t)$ is the bottom (top) of the adiabatic potential $E_+(t)(E_-(t))$, and

t_0 is the position at which the adiabatic energy difference becomes minimum. When the complex integral is annoying to evaluate, then the following formula can be used:

$$\sigma_{ZN} + i\delta_{ZN} = \frac{1}{\hbar}\left[\int_0^{t_0} E_+(t)dt - \int_0^{t_0} E_-(t)dt + \Delta\right] \tag{4.128}$$

with

$$\Delta = \frac{\sqrt{2}\pi}{4\sqrt{\alpha}} \frac{F_-^c + iF_+^c}{F_+^2 + F_-^2}, \tag{4.129}$$

where $F_\pm^c$ are the same as those given by Eqs. (4.28)–(4.31).

References

[1] H. Nakamura, *Nonadiabatic Transition: Concepts, Basic Theories and Applications*, World Scientific, 2002 (1st edition), 2012 (2nd edition).
[2] H. Nakamura, *Introduction to Nonadiabatic Dynamics*, World Scientific, 2019.
[3] I.-Y. Hsiao, Y. Teranishi and H. Nakamura, *Phys. Chem. Chem. Phys.* **26**, 3795 (2024).
[4] C. Zhu and H. Nakamura, *J. Chem. Phys.* **102**, 7448 (1995).
[5] C. Zhu and H. Nakamura, *J. Chem. Phys.* **109**, 4689 (1998).
[6] C. Zhu and H. Nakamura, *J. Chem. Phys.* **101**, 4855 (1994); Erratum, *J. Chem. Phys.* **108**, 7501 (1998).
[7] C. Zhu and H. Nakamura, *J. Chem. Phys.* **98**, 6208 (1993).
[8] C. Zhu and H. Nakamura, *J. Chem. Phys.* **101**, 10630 (1994).
[9] P. Oloyede, G.V. Mil'nikov and H. Nakamura, *J. Chem. Phys.* **124**, 144110 (2006).
[10] H. Nakamura, S. Nanbu, Y. Teranisi and A. Ohta, *Phys. Chem. Chem. Phys.* **18**, 11972 (2016).

Chapter 5

Periodic Sweeping (Chirping) of Laser Parameters

5.1 Basic Theory of Periodic Sweeping

The basic idea of periodic sweeping is explained with the use of a simple two-state system shown in Fig. 5.1 [1–4]. Here the two diabatic potential curves (dotted curves) are assumed to depend on the laser frequency $F(=\hbar\omega(t))$ with different signs of slope, namely, the original state is dressed up or down by one-photon energy $\hbar\omega$. However, this is just for clarity. One of the original diabatic states can be flat. This is the simple Landau–Zener–Stückerberg-type nonadiabatic transition. The off-diagonal element, i.e., the diabatic coupling, represents the laser-matter interaction. The adiabatic states (solid lines in Fig. 5.1), which are obtained by diagonalizing this diabatic Hamiltonian matrix, avoid crossing at the original diabatic crossing point. The nonadiabatic transition, i.e., a transition between these adiabatic curves, is induced by the time dependence of the matrix elements, and is known to occur locally in a small region of F near the avoided crossing point. If the diabatic states are linear functions of time t and the diabatic coupling is constant, then the nonadiabatic transition probability is given simply by the famous Landau–Zener formula, $p_{\mathrm{LZ}} = \exp[-2\pi V^2/\hbar\alpha]$, where V is the diabatic coupling strength. The parameter α is the slope difference of the diabatic states proportional to the speed of field sweeping at F_X, i.e., $\alpha \propto \dot{F}_X \equiv (dF/dt)_{F_X}$.

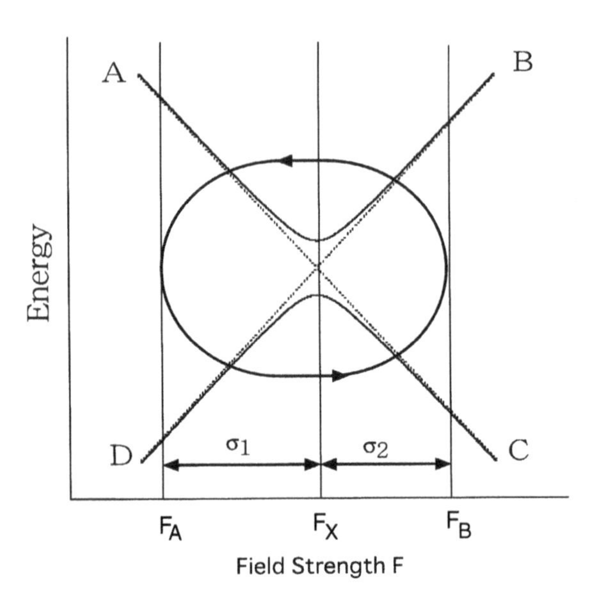

Figure 5.1: Schematic two diabatic (dotted lines) and two adiabatic (solid lines) potentials. The external field oscillates between F_A and F_B, striding the avoided crossing point F_X. σ_1 and σ_2 represent the phases accumulated in the designated ranges, and can be controlled by changing F_A and F_B. Reproduced with permission from [2].

The probability $p_{\rm LZ}$ takes a value in between 0 and 1, and becomes larger (smaller) with increasing (decreasing) $\dot{F}_X$. Thus, the simplest idea of controlling the nonadiabatic transition would be to change $p_{\rm LZ}$ as we wish by changing the field sweeping speed. The adiabatic rapid passage (ARP) is such an example. It is not easy at all, however, to directly control $p_{\rm LZ}$ in a wide range. Unreasonably rapid or slow sweeping is required.

Instead, if we change the field parameter F periodically over a certain range (F_A, F_B), we can control transitions between any pair of two states of A $\sim$ D in Fig. 5.1 as we desire. This can be realized by controlling phases associated with the field oscillation. Take, for instance, the transition from A to D in Fig. 5.1 by sweeping the field F for one period, i.e., F_A to F_B and back to F_A. Then the overall transition probability is given by

$$P_{AD} = 4p(1-p)\sin^2\psi, \tag{5.1}$$

where p is the nonadiabatic transition probability to which p_{LZ} gives a rough approximation, and ψ is the phase difference between the two possible paths A $\to$ C $\to$ D and A $\to$ B $\to$ D. The phase ψ can be controlled by changing the sweep range and thus the overall probability P_{AD} can be controlled to a desirable value between 0 and $4p(1-p)$. If p is equal to 0.5, we can control the transition as we wish. Even when p is not equal to 0.5, we can control P_{AD} to any desirable value as we wish by sweeping the field more than one period. That is to say, n-time periodical sweeping of the field creates 2^n multiple paths between the initial and final states and their phase interference can be controlled by changing the amplitude and the number n of the oscillation. If we want to control a transition from a state on the left side (A or D) to a state on the right (B or C) in Fig. 5.1, then we have to sweep the field $(n+1/2)$ periods. The necessary conditions for the control parameters, i.e., $\dot{F}_X$, (F_A, F_B), and n, can be formulated analytically and evaluated accurately with use of the new theory of time-dependent curve crossing problems developed in [5] based on the Zhu–Nakamura theory for the time-independent Landau-Zener-Stückerberg problems described in Chapter 4. This theory enables us to treat even the case $F_X \geq F_B$ in Fig. 5.1.

If p is very close to unity and we still want to realize an overall adiabatic passage, namely, a transition from A to A (or D to D) or from A to B (or D to C) in Fig. 5.1 with unit probability, a large number of oscillations (large n) is required as may be easily conjectured. The large number of oscillations is also required when p is very small and we want to realize an overall transition from A to C (or D to B) in Fig. 5.1 with unit probability. In this case, however, it would probably be possible to increase p substantially by simply reducing the field intensity to diminish the diabatic coupling strength. In the present case we sweep the field periodically and can find appropriate values of the control parameters explicitly with the help of analytical theory. Thus, the present method is quite efficient and actually can make the control perfect in principle.

The above-mentioned basic idea can, of course, be applied to a general multi-level system. Transitions at each avoided crossing can be controlled perfectly, in principle, and we can even specify a path from an initial state to any desired final state. One difficulty of the present method, however, arises when the density of avoided crossings is high and the oscillation amplitudes at each crossing overlap each other. In this case it is better to treat the bunch of avoided crossings as a whole, although its analytical treatment naturally becomes less straightforward. As mentioned above, the number of field oscillations becomes large, when the nonadiabatic transition probability p, for which we can use our new theory [5] which is much better than the Landau–Zener formula, is very close to unity and yet we want an overall adiabatic passage. A large number of oscillations could easily cause an experimental error and is better to be avoided. In this case we can think of using the non-crossing Rosen–Zener–Demkov-type nonadiabatic transition by employing the laser intensity as the adiabatic parameter F. Nonadiabatic transitions are also not necessarily restricted to the Landau–Zener–Stückerberg and the Rosen–Zener–Demkov cases, but can be any other general type such as the exponential models [6–8], although the availability of analytical theory is limited. We do not go into the details of these cases. The reader should refer to [2, 4].

As can be easily thought of from the above discussions, in the case of laser we may utilize both intensity and frequency as the adiabatic controlling parameters at the same time, and find an appropriate path of the adiabatic parameter F in the two-dimensional $(I,\ \omega)$ space.

The mathematical formulation of the above-mentioned idea is presented below. The most basic quantity is the following transition matrix I which describes a transition from F_A to F_B in Fig. 5.1 (this I should not be confused with the intensity I) [1]:

$$I = \begin{bmatrix} \sqrt{1-p}\,e^{i(\phi_S+\sigma_1/2+\sigma_2/2)} & -\sqrt{p}\,e^{i(\sigma_0-\sigma_1/2+\sigma_2/2)} \\ \sqrt{p}\,e^{-i(\sigma_0-\sigma_1/2+\sigma_2/2)} & \sqrt{1-p}\,e^{-i(\phi_S+\sigma_1/2+\sigma_2/2)} \end{bmatrix}. \quad (5.2)$$

The (i, j) element of this matrix provides the transition amplitude for the transition $j \to i$ $(i, j = 1, 2)$ when the field changes from F_A to F_B, where $i = 1$ (2) represents the lower (upper) adiabatic state. The backward transition from F_B to F_A can be described by the transpose of this matrix, I^T. Here p is the same as before, representing the nonadiabatic transition probability by one passage through the transition region, ϕ_S (Stokes phase) and σ_0 are the phase factors due to the nonadiabatic transition and their explicit expressions are given in Chapter 4. The phases σ_1 and σ_2 are the phase factors which describe, respectively, the adiabatic propagation in the region (F_A, F_X) and (F_X, F_B) in Fig. 5.1. They are given by

$$\sigma_1 = \int_{F_A}^{F_X} \Delta E(F) \frac{dt}{dF} dF = \int_{t_A}^{t_X} \Delta E(t) dt \tag{5.3}$$

and

$$\sigma_2 = \int_{F_X}^{F_B} \Delta E(F) \frac{dt}{dF} dF = \int_{t_X}^{t_B} \Delta E(t) dt, \tag{5.4}$$

where $\Delta E(F)$ is the adiabatic energy difference as a function of the adiabatic parameter F. The time t_α ($\alpha = A,\ B,\ X$) is the time at which $F(t_\alpha) = F_\alpha$ is satisfied. The final overall transition matrix after n periods of oscillation between F_A and F_B is expressed as

$$T_n = T^n, \tag{5.5}$$

where T is the transition matrix for one period of oscillation and is given by

$$T \equiv I^T I = \begin{bmatrix} \{p + (1-p)e^{2i\psi}\}e^{-i\sigma} & -2i\sqrt{p(1-p)}\sin\psi \\ -2i\sqrt{p(1-p)}\sin\psi & \{p + (1-p)e^{-2i\psi}\}e^{i\sigma} \end{bmatrix} \tag{5.6}$$

with $\psi \equiv \phi_S + \sigma_0 + \sigma_2$ and $\sigma \equiv 2\sigma_0 + \sigma_2 - \sigma_1$.[1] From Eqs. (5.5) and (5.6), the final transition probability after n periods of oscillation can be written as

$$P_{12}^{(n)} \equiv |(T_n)_{12}|^2 = |(T^n)_{12}|^2. \tag{5.7}$$

Using the Lagrange–Sylvester formula, we obtain

$$T_n = \frac{\lambda_+ \lambda_- (\lambda_-^{n-1} - \lambda_+^{n-1})}{\lambda_+ - \lambda_-} E + \frac{\lambda_+^n - \lambda_-^n}{\lambda_+ - \lambda_-} T, \tag{5.8}$$

[1]It should be pointed out here that Eqs. (1), (3), and (4), and the definitions of ψ and σ in [1] are not correct and should be replaced by the corresponding expressions given here.

where E is the unit matrix and $\lambda_\pm$ are the eigenvalues of T, which are given by

$$\lambda_\pm = e^{\pm i\xi}, \tag{5.9}$$

where

$$\cos\xi = (1-p)\cos(2\psi - \sigma) + p\cos(\sigma). \tag{5.10}$$

The unitarity of the matrix T requires ξ to be real. Equation (5.10) implies that the nonadiabatic transition probability p should satisfy

$$\frac{1-|\cos\xi|}{2} \le p \le \frac{1+|\cos\xi|}{2}. \tag{5.11}$$

From Eqs. (5.7) and (5.8), we can write $P_{12}^{(n)}$ as

$$P_{12}^{(n)} = 4\frac{\sin^2(n\xi)}{\sin^2\xi}p(1-p)\sin^2\psi. \tag{5.12}$$

Thus the conditions for $P_{12}^{(1)} = 0$ are $p = 0$, $p = 1$, or $\sin\psi = 0$, and the condition for $P_{12}^{(n)} = 0 (n > 1)$ is $\sin(n\xi) = 0$.

The conditions for $P_{12}^{(n)} = 1$ may be derived as follows. When $P_{12}^{(n)} = 1$ is satisfied, the diagonal terms of the transition matrix $T^{(n)}$ are zero and the transition matrix $T^{(2n)} = T^{(n)}T^{(n)}$ becomes diagonal ($P_{12}^{(2n)} = 0$). Thus the condition for $P_{12}^{(n)} = 1$ may be divided into the following two equations:

$$\sin^2(n\xi) = 1 \tag{5.13}$$

and

$$4p(1-p)\sin^2\psi = \sin^2\xi. \tag{5.14}$$

From Eqs. (5.11) and (5.12), we can estimate the number of oscillations n for a given p as the minimum integer which satisfies the following condition (see Fig. 5.2):

$$\sin^2\frac{\pi}{2n} \le 4p(1-p). \tag{5.15}$$

For a given n, Eq. (5.13) gives ξ, and ψ can be determined according to Eq. (5.14) for given ξ and p. Now, from the definitions of ψ and

σ, we can find the proper values of the two phase factors σ_1 and σ_2 to achieve $P_{12}^{(n)} = 1$, which can be adjusted by changing F_A and F_B.

Similar analysis can be done for n and half periods of oscillation. In this case, the final transition probability $P_{12}^{(n+1/2)}$ is given by

$$P_{12}^{(n+1/2)} \equiv |I(I^T I)^n|^2. \tag{5.16}$$

For both $P_{12}^{(n+1/2)} = 0$ and $P_{12}^{(n+1/2)} = 1$, the transition matrix after $2n+1$ period $T^{(2n+1)} = (I(I^T I)^n)^T I(I^T I)^n$ becomes diagonal. Thus we have

$$\sin((2n+1)\xi) = 0. \tag{5.17}$$

Substituting Eq. (5.17) into the condition $P_{12}^{(n+1/2)} = 0$ with the help of Eq. (5.16), we obtain

$$4(1-p)\sin^2(\psi - \sigma) = \frac{\sin^2 \xi}{\sin^2(n\xi)}. \tag{5.18}$$

On the other hand, for $P_{12}^{(n+1/2)} = 1$ we have

$$4p\sin^2(\psi - \sigma) = \frac{\sin^2 \xi}{\sin^2(n\xi)}. \tag{5.19}$$

In the case of both $P_{12}^{(n+1/2)} = 0$ and $P_{12}^{(n+1/2)} = 1$, n may be estimated as the minimum integer which satisfies the following condition (see Fig. 5.2):

$$\sin^2 \frac{\pi}{2n+1} \leq 4p(1-p). \tag{5.20}$$

The phases σ_1 and σ_2 are found in just the same way as in the case $P_{12}^{(n)} = 1$.

Now we can achieve the unit (or zero) overall transition probability for any p, ϕ_S and σ_0 by adjusting the number of oscillations n (or n and half) and the width of the oscillations F_A and F_B. In the following we consider the Landau–Zener–Stückelberg-type transition.

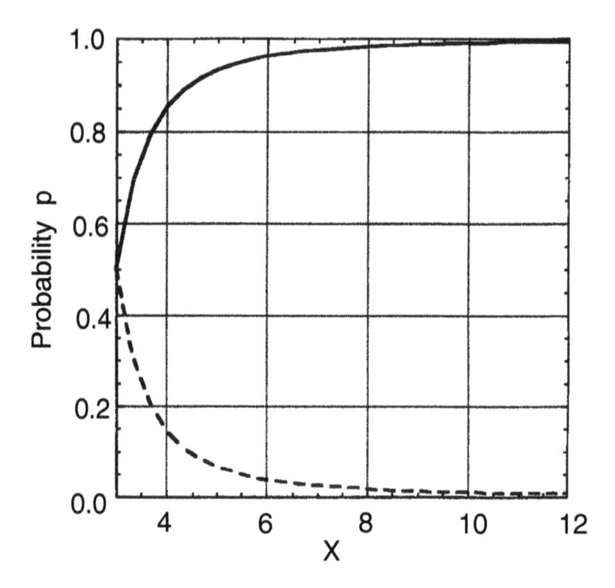

Figure 5.2: The solid and dotted curves represent $(1 + |\cos(\pi/(X))|)/2$ and $(1 - |\cos(\pi/(X))|)/2$, respectively. The range between these two curves represents the range of p to fulfill Eq. (5.11), namely the range in which complete control is achievable by n times of field oscillation. In the case of n period $X = 2n$, while $X = 2n+1$ for the n and half periods of oscillation of the field. Reproduced with permission from [2].

The Hamiltonian of the simplest linear potential model of Landau–Zener is given by

$$H_{\mathrm{LZ}} = \begin{bmatrix} \alpha_1 t & V \\ V & \alpha_2 t \end{bmatrix}, \tag{5.21}$$

where V is the constant diabatic coupling and t is the time. In the case of laser, this model corresponds to constant intensity and linear sweeping of the frequency. This model is good enough to explain qualitative features of the control scheme. The nonadiabatic transition occurs at $t = 0$ with the transition probability p given by

$$p_{\mathrm{LZ}} = \exp\left(-2\pi \frac{V^2}{\hbar\alpha}\right) = \exp\left(-2\pi \frac{I\epsilon^2}{\hbar^2|l - m|\dot{\omega}}\right), \tag{5.22}$$

where $\alpha \equiv |\alpha_1 - \alpha_2|$, I and $\dot{\omega}$ are the laser intensity and the sweeping speed of the frequency at the avoided crossing, respectively, l and

m (can be negative) are the photon numbers, and ϵ is the dipole matrix element. The conventional adiabatic passage requires large laser intensity and small sweeping speed to make $p_{\rm LZ}$ very small. For instance, the adiabatic passage with $p_{\rm LZ} \leq 0.001$ requires

$$1.0994\hbar\dot{\omega}/\epsilon \leq I. \tag{5.23}$$

For the values $\epsilon = 1.0$ (Åe) and $\dot{\omega} = 0.5$ (cm^{-1}/ps), I must be larger than 1.0 (TW/cm^2).

In order to accomplish the passage in a reasonably short time scale with relatively large $\dot{\omega}$, a large intensity I or a large number of oscillations is required. If $p_{\rm LZ} = 0.5$ can be attained without difficulty, then one period of oscillation enables us to achieve exactly zero or unit final transition probability. The required intensity in this case is given by

$$I = 0.1103\hbar\dot{\omega}/\epsilon. \tag{5.24}$$

Namely, one period of oscillation requires the intensity to be one order smaller compared to the case of one passage for the same ϵ and $\dot{\omega}$. Furthermore, ten periods of oscillation require the following condition (see Eq. (5.12)):

$$0.982 \times 10^{-3}\hbar\dot{\omega}/\epsilon \leq I \leq 0.810\hbar\dot{\omega}/\epsilon. \tag{5.25}$$

This requires only one-thousandth of the intensity required in the case of one passage. Many periods of oscillation, however, require high accuracy of p and phases as discussed above. When p is large, it is sensitive to the error in the exponent which is proportional to the intensity. In the case of adiabatic passage, on the other hand, p is relatively stable against the error in the exponent, since the exponent is large. When $p \simeq 0.5$, about 15% of error in the exponent yields the fluctuation in the range $0.45 < p_{\rm LZ} < 0.65$.

So far, our discussion is based on the simple model Eq. (5.21). For finding the actual parameters, however, it is much better to use the sophisticated theory developed in [5], because the theory is applicable to general functionality of ω, even if the two diabatic potentials touch each other or avoid crossing.

5.2 Selective and Complete Excitation of Energy Levels

5.2.1 *In the Case of Three and Four Levels*

5.2.1.1 Three-level (1 + 2) Case

First, let us consider a three-level system shown in Fig. 5.3 [3, 4, 9, 10]. The energy separation ω_{23} is assumed to be much smaller than the separation between the ground and excited states, namely, $\omega_{12} \gg \omega_{23}$, where $\omega_{ij} = (E_j - E_i)/\hbar$. Since the applied laser frequency ω is close to ω_{12}, the transition between the levels $|2\rangle$ and $|3\rangle$ are negligible and the Floquet Hamiltonian is expressed as

$$H_{Floquet} = \begin{pmatrix} E_1 + \hbar\omega(t) & -\mu_{12}\epsilon(t)/2 & -\mu_{13}\epsilon(t)/2 \\ -\mu_{12}\epsilon(t)/2 & E_2 & 0 \\ -\mu_{13}\epsilon(t)/2 & 0 & E_3 \end{pmatrix}, \tag{5.26}$$

where μ_{ij} is the transition dipole moment between $|i\rangle$ and $|j\rangle$ and $\epsilon(t)$ is an envelope function of the laser field. When the frequency is swept from ω_1 to ω_2, the corresponding transition matrix is given by

$$T_{\omega_1 \to \omega_2} = \Sigma_3 I_2 \Sigma_2 I_1 \Sigma_1, \tag{5.27}$$

where Σ_j represents the adiabatic propagation from X_{j-1} to X_j, and I_j represents the nonadiabatic transition amplitude at the crossing X_j. They are explicitly given by

$$(\Sigma_j)_{pq} = \exp[-i\sigma_j^{(p)}]\delta_{pq}, \tag{5.28}$$

$$\sigma_j^{(k)} = \frac{1}{\hbar}\int_{X_{j-1}}^{X_j} E_k^{(a)}(\omega)dt, \tag{5.29}$$

$$I_1 = \begin{pmatrix} \sqrt{1-p_1}\exp[i\phi_1] & \sqrt{p_1}\exp[i\psi_1] & 0 \\ -\sqrt{p_1}\exp[-i\psi_1] & \sqrt{1-p_1}\exp[-i\phi_1] & 0 \\ 0 & 0 & 1 \end{pmatrix} \tag{5.30}$$

$$I_2 = \begin{pmatrix} 1 & 0 & 0 \\ 0 & \sqrt{1-p_2}\exp[i\phi_2] & \sqrt{p_2}\exp[i\psi_2] \\ 0 & -\sqrt{p_2}\exp[-i\psi_2] & \sqrt{1-p_2}\exp[-i\phi_2] \end{pmatrix}. \tag{5.31}$$

Here, p_j denotes the nonadiabatic transition probability for one passage of the avoided crossing point X_j, ϕ_j and ψ_j are the dynamical

(a)

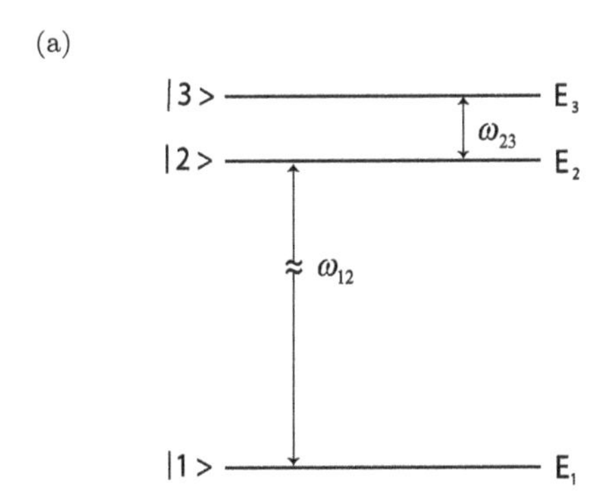

(b)

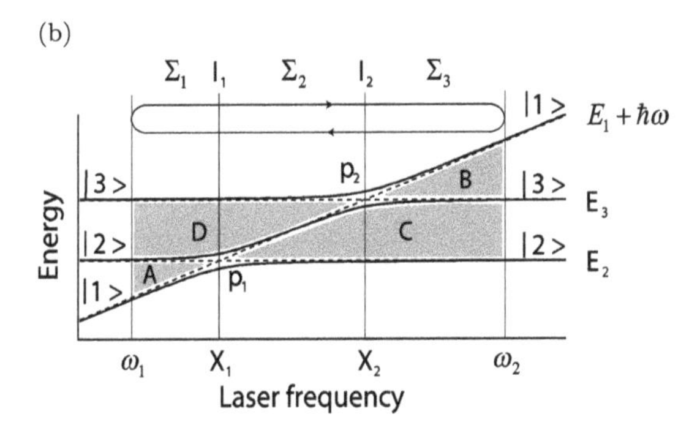

Figure 5.3: (a) Schematic level structure of a three-level model. (b) Floquet diagram corresponding to the three-level model in (a) as a function of laser frequency. Reproduced with permission from [10].

phases due to the nonadiabatic transition at X_j, $E_k^{(a)}$ is the k-th adiabatic Floquet state, $X_0 = \omega_1$ and $X_3 = \omega_2$. The total transition amplitude can be expressed as

$$T_{\omega_1 \to \omega_2} = \begin{pmatrix} \sqrt{1-p_1} & \sqrt{p_1}\exp[-iA] & 0 \\ -\sqrt{p_1(1-p_2)}\exp[-iC] & \sqrt{(1-p_1)(1-p_2)}\exp[-i(A+C)] & \sqrt{p_2}\exp[-i(A+C+D)] \\ \sqrt{p_1 p_2}\exp[-i(B+C)] & -\sqrt{(1-p_1)p_2}\exp[-i(A+B+C)] & \sqrt{1-p_2}\exp[-i(A+B+C+D)] \end{pmatrix}, \tag{5.32}$$

where

$$A = \phi_1 - \psi_1 + \Delta\sigma_1^{(2,1)}, \tag{5.33}$$

$$B = \phi_2 + \psi_2 + \Delta\sigma_3^{(3,2)}, \tag{5.34}$$

$$C = \phi_1 + \psi_1 - \phi_2 + \Delta\sigma_2^{(2,1)} + \Delta\sigma_3^{(2,1)}, \tag{5.35}$$

$$D = -\phi_1 + \phi_2 - \psi_2 + \Delta\sigma_1^{(3,2)} + \Delta\sigma_2^{(3,2)}, \tag{5.36}$$

$$\Delta\sigma_j^{(k,l)} = \sigma_j^{(k)} - \sigma_j^{(l)}. \tag{5.37}$$

These phases $A \sim D$ roughly correspond to the areas shown in Fig. 5.3(b). The overall transition matrix for one period of oscillation is given by

$$T^{(n=1)} = (T_{\omega_1 \to \omega_2})^T T_{\omega_1 \to \omega_2} \tag{5.38}$$

and the transition probability from $|1\rangle$ to $|2\rangle$ is explicitly given by

$$P_{12}^{(n=1)} = p_1(1 - p_1)|\exp[2iC] - 1 + p_2(1 - \exp[-2iB])|^2. \tag{5.39}$$

The condition for the complete excitation to $|2\rangle(P_{12}^{(n=1)} = 1)$ is expressed as

$$p_1 = 1/2, \quad B = m\pi \quad \text{and} \quad C = (n + 1/2)\pi \quad (m, n = \text{integer}). \tag{5.40}$$

The physical meaning of $B = m\pi$ is that no bifurcation into the diabatic state $|3\rangle$ occurs at X_2 on the second half of the sweep whatever the probability p_2 is. The conditions of p_1 and C guarantee that the interference between $|1\rangle$ and $|2\rangle$ at X_1 on the way back leads to the complete excitation to $|2\rangle$. The complete and selective excitation to $|3\rangle$ can be achieved by one period of sweeping, if we start from ω_2. The condition is given by

$$p_2 = 1/2, \quad A = m\pi \quad \text{and} \quad D = (n + 1/2)\pi \quad (m, n = \text{integer}). \tag{5.41}$$

Numerical examples are shown in Fig. 5.4. The parameters used are

$$\omega_{12} = 500 \text{ cm}^{-1}, \quad \omega_{23} = 10 \text{ cm}^{-1}, \quad \mu_{12} = \mu_{13} = 1.0 \text{ a.u.} \tag{5.42}$$

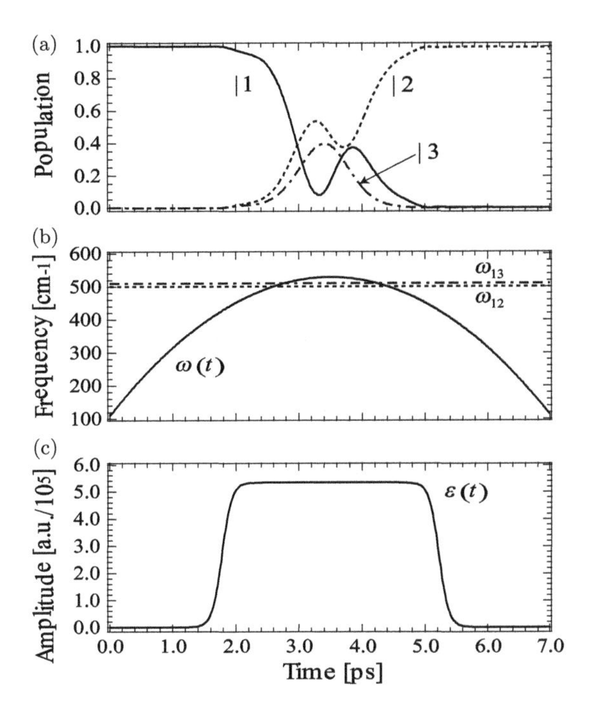

Figure 5.4: Complete excitation from $|1\rangle$ to $|2\rangle$ by one period of chirping in the case of three-level model in Fig. 5.3. (a) Time variation of the population, (b) time variation of laser frequency, and (c) envelope of the laser pulse. Reproduced with permission from [10].

The laser frequency is swept quadratically as a function of time as

$$\omega(t) = -a(t - t_0)^2 + b - E_1, \tag{5.43}$$

where

$$a = 8(V_{13})^3 \alpha_f / \hbar^2, \quad b = E_3 - 2V_{13}\beta_f, \quad V_{13} = -\mu_{13}E_0/2 \tag{5.44}$$

with E_0 being the laser amplitude at the peak. The laser pulse shape is taken as

$$\epsilon(t) = \begin{cases} E_0[1 + \tanh(\beta_e(t - t_{0e}))]/2 & \text{for } t \leq t_0 \\ E_0[1 - \tanh(\beta_e(t - t_{1e}))]/2 & \text{for } t > t_0. \end{cases} \tag{5.45}$$

The parameters α_f, β_f and the peak intensity are determined to be 0.6005, 1.58142, and 0.1 GW/cm^2 from the condition given above and the Zhu-Nakamura formulas. The other parameters of laser field are $t_0 = 3.5$ ps, $t_{0e} = 1.7775$ ps, $t_{1e} = 2t_0 - t_{0e}$, and $\beta_e = 6.515$ ps^{-1}. As is seen from Fig. 5.4, the transition time is $\sim$3 ps, which is very close to the time determined from the uncertainty principle, $\delta t = 2\pi/\Delta E \simeq 3.3$ ps. Thus the present scheme gives the shortest possible time of transition. In the case of ARP method selective excitation is possible, but it takes quite a long time and complete excitation is not possible. The selective excitation with the excitation probability $\sim$0.99 is attained by using the linear chirping, $\omega(t) = \omega_{12} + c(t - 20\text{ ps})$ with $c = 8.816$ cm^{-1}/ps, and the laser intensity 0.1 GW/cm^2. It takes quite a long time of $\sim$20 ps. In the case of π-pulse, the selective and complete excitation is, of course, achievable, but the pulse duration becomes longer at $\sim$14 ps. The details are not shown here (see [9]).

5.2.1.2 Four-level (1 + 3) Case

Next, let us consider the four-level system, in which three higher levels $|2\rangle, |3\rangle$ and $|4\rangle$ are close to each other, i.e., $\omega_{12} \gg \omega_{23}, \omega_{34}$ [10]. When the laser frequency is swept from ω_1 to ω_2, as in the three-level case, the transition matrix in the adiabatic state representation is expressed as

$$T_{\omega_1 \to \omega_2} = \Sigma_4 I_3 \Sigma_3 I_2 \Sigma_2 I_1 \Sigma_1, \tag{5.46}$$

where Σ_j and I_j represent, as before, the adiabatic propagation from X_{j-1} to X_j and the nonadiabatic transition matrix (I-matrix) at X_j. The transition matrix $T_{\omega_1 \to \omega_2}$ can be explicitly expressed as

$$T_{\omega_1 \to \omega_2} = \begin{pmatrix} \sqrt{1-p_1} & \sqrt{p_1}e^{-iA} & 0 & 0 \\ -\sqrt{p_1(1-p_2)}e^{-iC} & \sqrt{(1-p_1)(1-p_2)}e^{-i(A+C)} & \sqrt{p_2}e^{-i(A+C+D)} & 0 \\ \sqrt{p_1p_2(1-p_3)}e^{-i(C+E)} & -\sqrt{(1-p_1)p_2(1-p_3)}e^{-i(A+C+E)} & \sqrt{(1-p_2)(1-p_3)}e^{-i(A+C+D+E)} & \sqrt{p_3}e^{-i(A+C+D+E+F)} \\ -\sqrt{p_1p_2p_3}e^{-i(B+C+E)} & \sqrt{(1-p_1)p_2p_3}e^{-i(A+B+C+E)} & -\sqrt{(1-p_2)p_3}e^{-i(A+B+C+D+E)} & \sqrt{1-p_3}e^{-i(A+B+C+D+E+F)} \end{pmatrix}, \tag{5.47}$$

where

$$A = \phi_1 - \psi_1 + \Delta\sigma_1^{(2,1)}, \quad B = \phi_3 + \psi_3 + \Delta\sigma_4^{(4,3)}, \tag{5.48}$$

$$C = \phi_1 + \psi_1 - \phi_2 + \Delta\sigma_2^{(2,1)} + \Delta\sigma_3^{(2,1)} + \Delta\sigma_4^{(2,1)}, \tag{5.49}$$

$$D = -\phi_1 + \phi_2 - \psi_2 + \Delta\sigma_1^{(3,2)}, \tag{5.50}$$

$$E = \phi_2 + \psi_2 - \phi_3 + \Delta\sigma_3^{(3,2)} + \Delta\sigma_4^{(3,2)}, \tag{5.51}$$

$$F = -\phi_2 + \phi_3 - \psi_3 + \Delta\sigma_1^{(4,3)} + \Delta\sigma_2^{(4,3)} + \Delta_3^{(4,3)}. \tag{5.52}$$

The definitions of p_j, ϕ_j, ψ_j and $\sigma_j^{(k,l)}$ are the same as before.

The transition matrix $T^{(1)}$ for one period sweeping is defined by Eq. (5.38). The elements for the transition $1 \to j (j = 2-4)$ are given as follows:

$$[T^{(1)}]_{21} = \sqrt{p_1(1-p_1)}e^{-iA}X, \tag{5.53}$$

$$[T^{(1)}]_{31} = -\sqrt{p_1 p_2(1-p_2)}e^{-i(A+2C+D)}Y, \tag{5.54}$$

$$[T^{(1)}]_{41} = \sqrt{p_1 p_2 p_3(1-p_3)}e^{-i(A+2C+D+2E+F)}Z, \tag{5.55}$$

where

$$X = 1 - e^{-2iC} + p_2 e^{-2iC}Y, \quad Y = 1 - e^{-2iE} + p_3 e^{-2iE}Z,$$
$$Z = 1 - e^{-2iB}. \tag{5.56}$$

From the conditions, $|(T^{(1)})_{21}| = 1$, $|(T^{(1)})_{31}| = 0$ and $|(T^{(1)})_{41}| = 0$, the condition of the complete excitation from $|1\rangle$ to $|2\rangle$ is obtained as

$$p_1 = 1/2, \quad B = l\pi, \quad E = m\pi \text{ and } C = (n+1/2)\pi \quad (l, m, n\text{: integer}). \tag{5.57}$$

The complete excitation from $|1\rangle$ to $|2\rangle$ is considered next. It turns out that at least one and a half period of sweeping is needed in this case. The overall transition matrix $T^{(1+1/2)}$ is given by $T^{(1+1/2)} = T_{\omega_1 \to \omega_2} \cdot T^{(1)}$ and its $1 \to j (j = 1-4)$ elements are explicitly expressed as

$$(T^{(1+1/2)})_{11} = \sqrt{1-p_1}(1 - p_1X + p_1e^{-2iA}X), \tag{5.58}$$

$$(T^{(1+1/2)})_{21} = \sqrt{p_1(1-p_2)}e^{-iC}\{-(1-p_1X) + (1-p_1)e^{-2iA}X - p_2Ye^{-2i(A+C+D)}\}, \tag{5.59}$$

$$(T^{(1+1/2)})_{31} = \sqrt{p_1p_2(1-p_3)}e^{-i(C+E)}\{1 - p_1X - (1-p_1)e^{-2iA}X - (1-p_2)Ye^{-2i(A+C+D)} + p_3Ze^{-2i(A+C+D+E+F)}\}, \tag{5.60}$$

$$(T^{(1+1/2)})_{41} = \sqrt{p_1p_2p_3}e^{-i(B+C+E)}\{-(1-p_1X) + (1-p_1)e^{-2iA}X + (1-p_2)Ye^{-2i(A+C+D)} + (1-p_3)Ze^{-2i(A+C+D+E+F)}\}. \tag{5.61}$$

From the conditions, $|(T^{(1+1/2)})_{21}| = 1$, $|(T^{(1+1/2)})_{11}| = |(T^{(1+1/2)})_{31}| = |(T^{(1+1/2)})_{41}| = 0$, and $|(T^{(1)})_{41}| = 0$, the following conditions are obtained:

$$4p_1(1-p_2)\sin^2 E = 1, \quad B = k\pi, \quad D = l\pi, \quad A + C = m\pi, \quad C + E = n\pi \tag{5.62}$$

or

$$4p_1(1-p_2)\sin^2 E = 1, \quad B = k\pi, \quad A + C + D + E = (l+1/2)\pi, \quad C + D = (m+1/2)\pi, \tag{5.63}$$

where k, l, m, n are integers.

The conditions for the complete excitation $|1\rangle \to |3\rangle$ are a bit more complicated. One particular solution is found to be as follows:

$$A = (l+1/2)\pi, \quad B = (m+1/2)\pi, \quad C = n\pi, \quad D = p\pi, \quad E = (q+1/2)\pi, \quad F = r\pi, \quad X = p_2, \quad Y = 1, \quad Z = 2, \quad p_1p_2 = 1/2 \quad \text{and} \quad p_3 = 1/2, \tag{5.64}$$

where l, m, n, p, q, r are integers.

Numerical examples are shown in Figs. 5.5 and 5.6. The parameters used are as follows: $\omega_{12} = 500\ \text{cm}^{-1}, \omega_{34} = 10\ \text{cm}^{-1}, \mu_{12} = \mu_{13} = \mu 14 = 1.0$ a.u. Here we present the complete excitation of $|3\rangle$. The result of one and a half period of sweeping is shown in Fig. 5.5. The laser frequency is linearly chirped as follows:

$$\omega(t) = \begin{cases} \omega_{14} + c(t - t_0) & (t \leq t_1) \\ \omega_{14} - c(t - t_2) & (t_1 < t \leq t_3) \\ \omega_{14} + c(t - t_4) & (t > t_3), \end{cases} \tag{5.65}$$

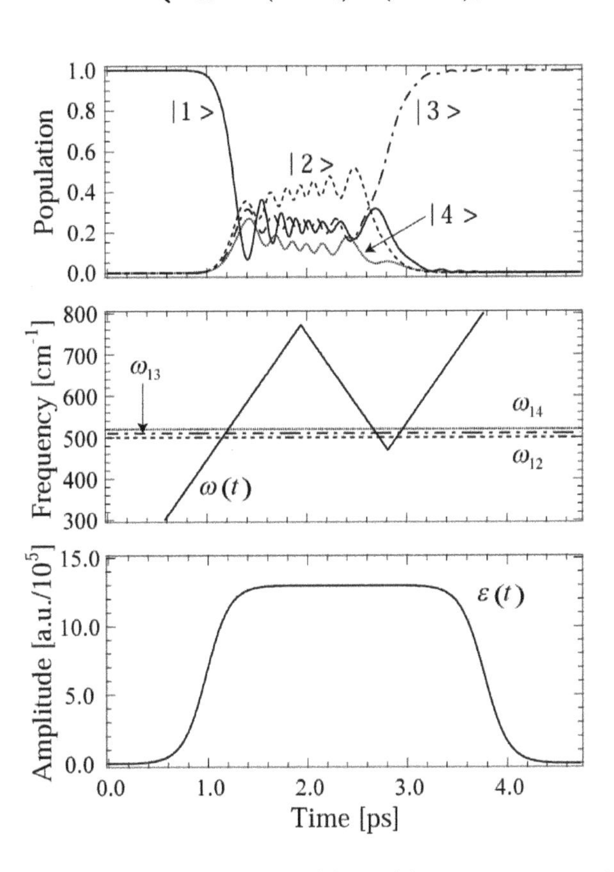

Figure 5.5: Complete excitation from $|1\rangle$ to $|3\rangle$ by one and a half period of frequency chirping in the case of four-level model. Time variations of the population (upper part), laser frequency (middle part) and laser envelope (bottom part) are shown. Reproduced with permission from [10].

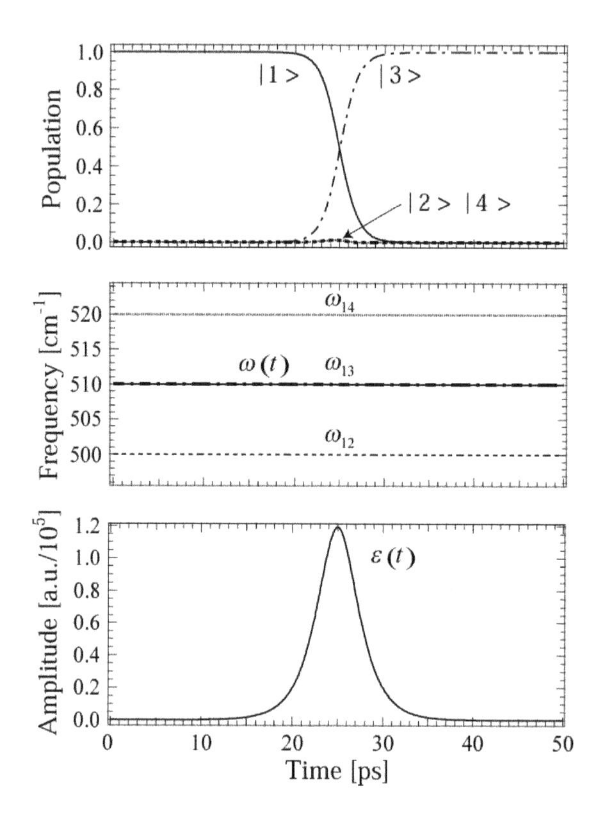

Figure 5.6: Excitation from $|1\rangle$ to $|3\rangle$ by π-pulse. Complete excitation is possible, but the pulse duration is long. Reproduced with permission from [10].

where $t_1 = t_0 + \Delta t_1$, $t_2 = t_0 + 2\Delta t_1$, $t_3 = t_2 + \omega_{24}/c + \Delta t_2$, $t_4 = 2t_3 - t_2$, $c = 346.6$ cm^{-1}/ps, $\Delta t_1 = 726.3$ fs, $\Delta t_2 = 90.08$ fs and the peak intensity is 0.5917 GW/cm^2. The pulse shape is again a combination of hyperbolic-tangent functions defined above with $t_{0e} = 981.2$ fs and $\beta_e = 4.257$ ps^{-1}. The transition time is about 3 ps and this is close to the time limit determined by the uncertainty principle. Fig. 5.6 shows the result of π-pulse with a long pulse duration. The parameters are $t_0 = 25$ ps, the peak intensity $= 0.005$ GW/cm^2 and $\beta_e = 0.493386$ ps^{-1}. In order to achieve complete excitation, the pulse duration is much longer than the present one and a half chirping case. Needless to say, the ARP takes a much longer time, since the two ARP processes, $|1\rangle$ to $|2\rangle$ and $|2\rangle$ to $|3\rangle$, are needed.

5.2.2 *Ring-Puckering Isomerization of Tri-methylenimine*

Here, we consider the laser-induced ring-puckering isomerization of tri-methylenimine (see Fig. 5.7) which was discussed by Sugawara and Fujimura in [11]. This problem may be reduced to a one-dimensional double-well problem (see Fig. 5.8), in which the left (right) well corresponds to the isomer A (B), and the isomerization from A to B occurs through tunneling from the left well to the right well. We try to control this isomerization by the quadratic chirping of laser pulses. All the parameters to determine the potential system are taken from [11].

The Floquet state diagram as a function of laser frequency ω [cm^{-1}] with constant intensity ($I = 0.1$ [$\mathrm{TW/cm}^2$]) is shown in Fig. 5.9. There appears to be a lot of avoided crossings, where the energy gap is proportional to the laser intensity I and the square of the transition dipole moment between the corresponding two states of slope ± 1. We can treat each avoided crossing separately unless the laser intensity is extremely strong and avoided crossings overlap with each other. The curves of slope ± 1 correspond to the one-photon absorption or emission, the curves of slope ± 2 are the two-photon absorption/emission processes and so on.

For a transition between two states with a small transition moment, a larger intensity or a smaller sweep velocity is required. This means that the direct isomerization from $|0\rangle$ to $|1\rangle$ requires very large intensity or a very long transition time (very slow sweeping).

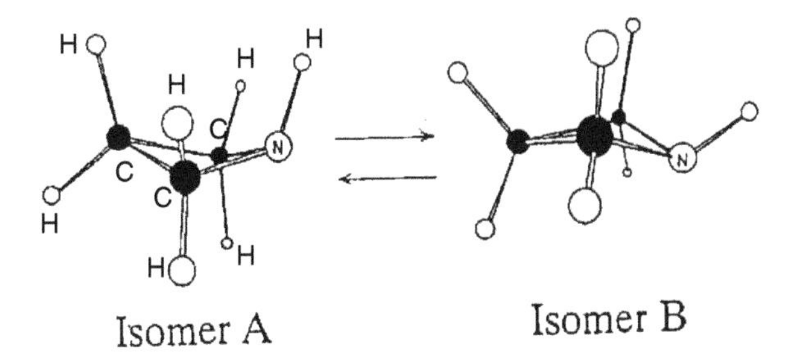

Figure 5.7: Molecular structure of tri-methylenimine, $(CH_2)_3NH$, and its puckering isomerization (A $\Longleftrightarrow$ B).

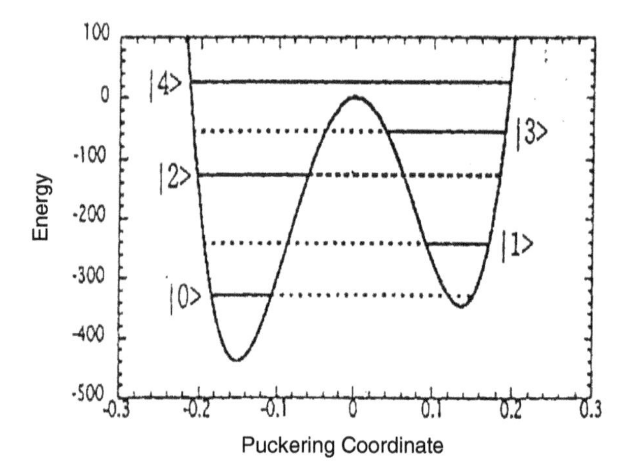

Figure 5.8: Double-well potential model of tri-methylenimine. The origin of potential energy is taken at the barrier top. Coordinate is in Å and energy is in cm^{-1}. Reproduced with permission from [2].

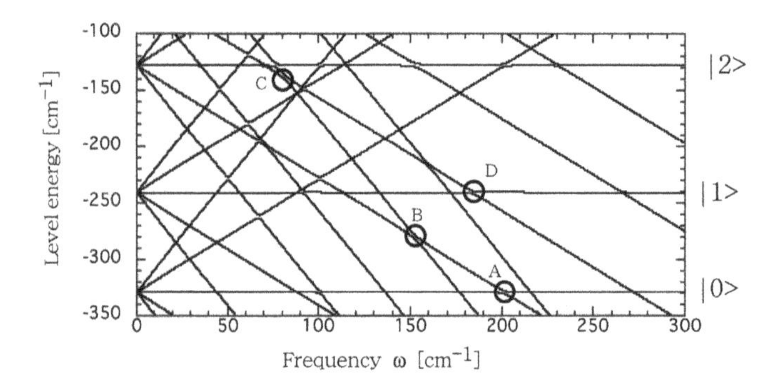

Figure 5.9: Vibrational energy levels (Floquet states) of tri-methylenimine as a function of laser frequency. Reproduced with permission from [2].

For controlling the isomerization, it is better to use an indirect process which is composed of the transitions of relatively large transition moments [11]. It should be noted that the square of the transition moment for the direct process $|0\rangle \to |1\rangle$ is about four orders of magnitude smaller than that for $|0\rangle \to |2\rangle$. Fig. 5.10 shows an example of such indirect isomerization [2, 3]: $|0\rangle \to |2\rangle \to |4\rangle \to |3\rangle \to |1\rangle$, where four pulses are applied corresponding to these

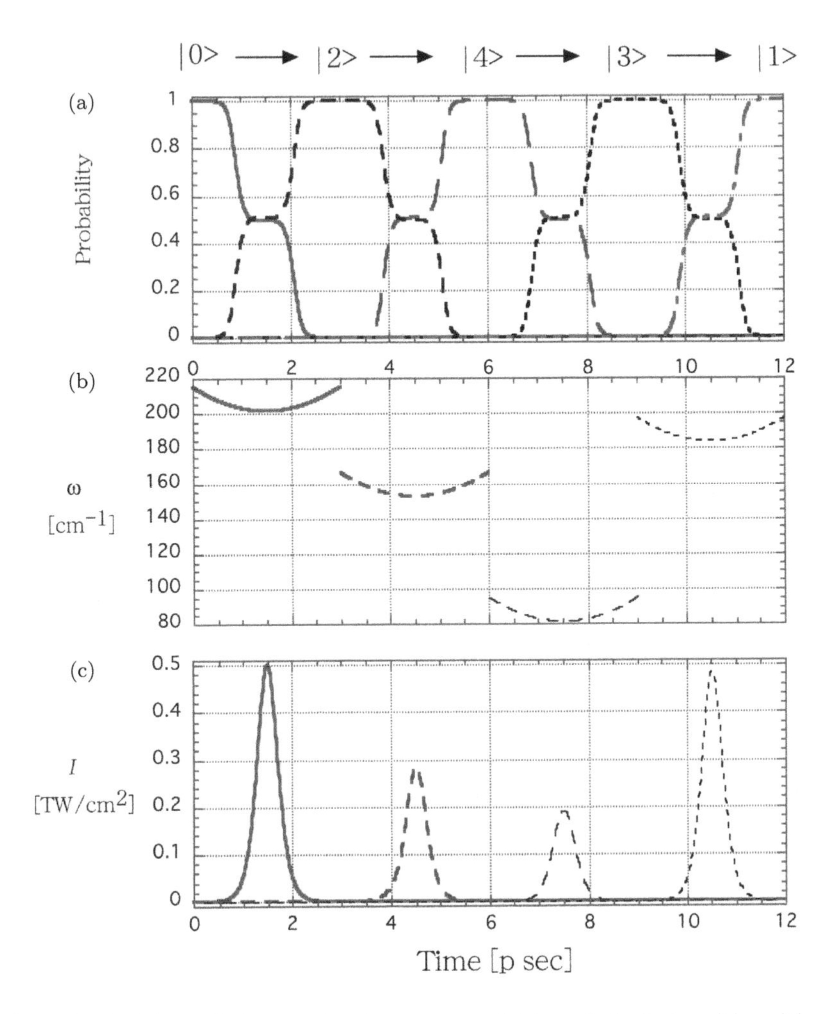

Figure 5.10: Controlled isomerization process of tri-methylenimine, $|0\rangle \to |2\rangle \to |4\rangle \to |3\rangle \to |1\rangle$. The corresponding avoided crossings used are designated as A-D in Fig. 5.9. The corresponding resonance frequencies are 202.6, 153.6, 82.6 and 185.2 cm^{-1}. At each transition complete transition is attained. Reproduced with permission from [2].

four transitions. That is to say, the first pulse achieves the complete transition $|0\rangle \to |2\rangle$, and the second one does $|2\rangle \to |4\rangle$, and so on. The corresponding avoided crossings are designated as A–D in Fig. 5.9. To obtain Fig. 5.10, 12 Floquet states are taken into account in numerical computations. Here, we have used exactly the same shape of $\omega(t) = at^2 + b$ at four avoided crossings. The pulse shapes are proportional to sechyp^2.

5.2.3 *Selective Excitation of Fine Structure States of* K *and* Cs *Atoms*

Let us next consider the real atoms K and Cs, and demonstrate the selective and complete excitation to one of the close-lying fine structure states of these atoms [3, 4, 12]. To ensure that no leak excitation occurs to other excited states, time-dependent coupled Schrödinger equations are solved by expanding the total wave function in terms of 10(18) unperturbed atomic eigenstates for K(Cs). The convergence of numerical results is confirmed.

5.2.3.1 In the case of K

Fig. 5.11 shows the time variation of the populations of the ground 4S (solid line), and excited $4P_{1/2}$ (dashed line) and $4P_{3/2}$ (dotted line) states of K [3, 12]. The laser intensity is 0.36 GW/cm^2 with a quadratic chirping shown in Fig. 5.11(b). The chirping is concave down, which makes it possible to completely suppress the transition to the upper excited state and realize the selective excitation to the lower excited state ($4P_{1/2}$) with 100% efficiency. The excitation energies of the K atom from the ground state to $4P_{1/2}$ and $4P_{3/2}$ are 12 985.17 and 13 042.88 cm^{-1}, respectively, and thus the energy splitting between them is equal to $\Delta E = 57.7$ cm^{-1}. This energy splitting corresponds to the time $\Delta T = 2\pi/\Delta E \simeq 577$ fs estimated from the uncertainty principle. As is seen from Fig. 5.11(a), the quadratic chirping scheme enables us to achieve the complete and selective excitation within 600 fs that is very close to ΔT. It should be emphasized again that our scheme here realizes selective excitation with unit probability as quickly as the uncertainty principle limit. Selective excitation to the upper ($4P_{3/2}$) state can be achieved with use of concave-up chirping instead of concave-down, as is shown in Fig. 5.12. The laser intensity is 0.13 GW/cm^2, which is about three times smaller than the previous case. This is because the transition dipole moment between the ground and the upper excited state is larger than that between the ground and the lower excited state.

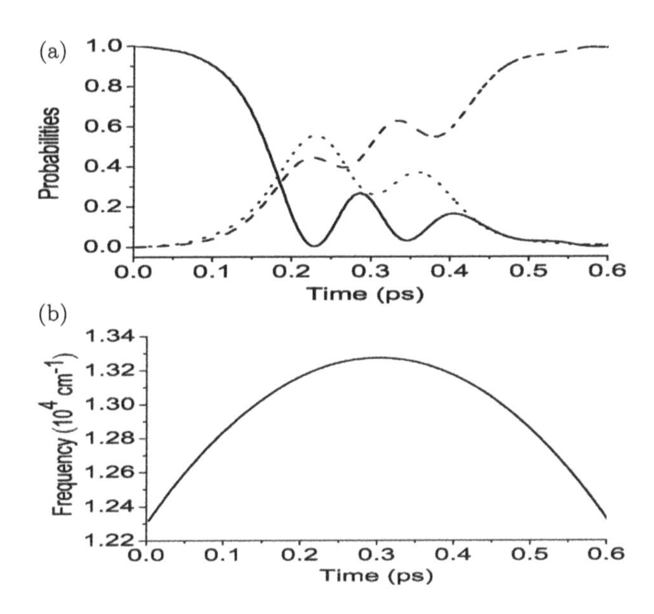

Figure 5.11: (a) Time variation of the populations of the ground 4S (solid line), excited $4P_{1/2}$ (dashed line) and $4P_{3/2}$ (dotted line) states of K atom. (b) Time-dependent frequency to induce the selective excitation to $4P_{1/2}$. Reproduced with permission from [12].

5.2.3.2 In the case of Cs

The next example is the selective excitation of Cs atom from its ground state (6S) to any one of the excited states ($7D_{3/2}$ and $7D_{5/2}$) [3, 4, 12]. The excitation energies are 26 047.86 and 26 068.83 cm^{-1}, respectively, which are roughly two times larger than those of the previous K case. Thus, considering the symmetry of the excited states also, we have to employ two-photon processes. As long as the transitions (or avoided crossings) are separated in time, there is no difficulty in achieving the selection. Fig. 5.13(a) shows the time variation of populations of the ground 6S (solid line) and excited $7D_{3/2}$ (dashed line) and $7D_{5/2}$ (dotted line) states of Cs. The laser intensity is 4.8 GW/cm^2, and the frequency chirping is depicted in Fig. 5.13(b). As is clearly seen in Fig. 5.13(a), the complete and selective excitation to the $7D_{3/2}$ (lower) state is achieved by the concave-down quadratic chirping. Here the frequency is chirped around half of the resonance, namely, ~13 000 cm^{-1}, because the

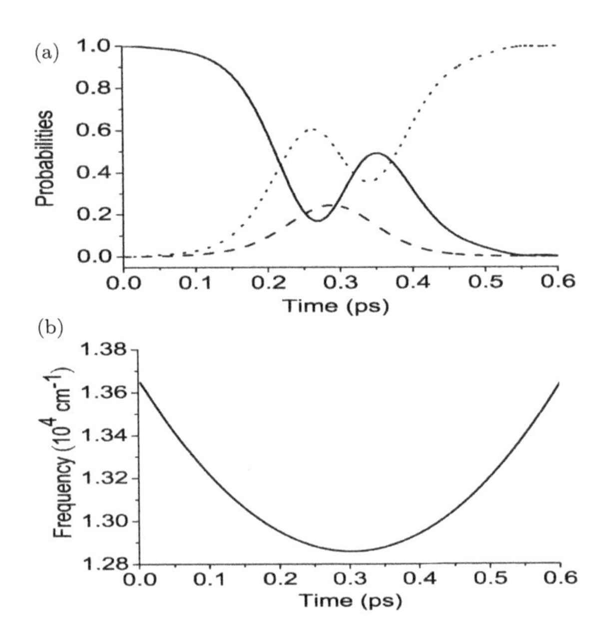

Figure 5.12: (a) Time variation of the populations of the ground 4S (solid line), excited $4P_{1/2}$ (dashed line) and $4P_{3/2}$ (dotted line) states of K atom. (b) Time-dependent frequency to induce the selective excitation to $4P_{3/2}$. Reproduced with permission from [12].

transitions take place by the two-photon absorption. The control is accomplished in 3 ps, which is roughly twice as long as the uncertainty principle limit given by $\Delta T = 2\pi/\Delta E \sim 1.6$ ps. Although our control scheme works well no matter whether it is one- or two-photon process, it is difficult to achieve complete and selective excitation in the case of the two-photon process as quickly as the uncertainty principle limit allows, as in the case of K. The reason is as follows. To induce two-photon transitions, a larger laser intensity is required, which may cause off-resonance transitions. In our case of Cs, the off-resonance excitation to 6P states becomes non-negligible when the laser intensity exceeds 5 GW/cm^2. Since the present analytical formulation given above is for three-level problems, the laser intensity should be small to suppress these undesired transitions. To satisfy the condition $p = 1/2$ with a small laser intensity, a long transition time is necessary. This is why long control time is required.

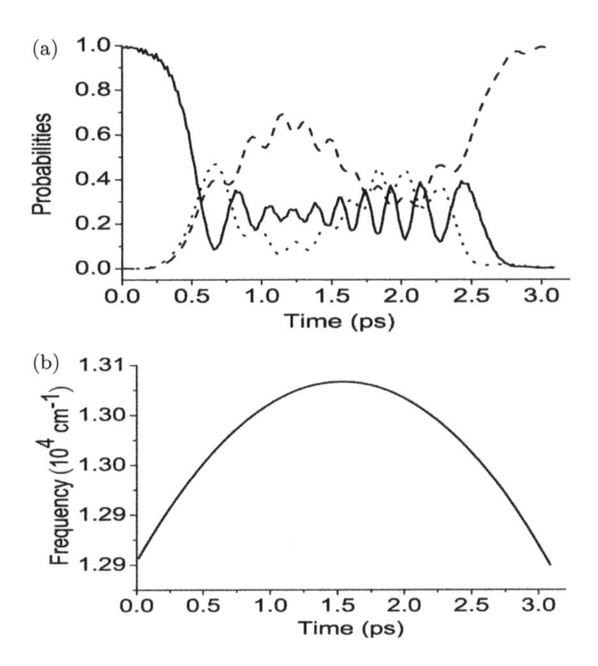

Figure 5.13: (a) Time variation of the populations of the ground 6S (solid line), excited $7D_{3/2}$ (dashed line) and $7D_{5/2}$ (dotted line) states of Cs atom. (b) Time-dependent frequency to induce the selective excitation to $7D_{3/2}$. Reproduced with permission from [12].

5.3 Excitation and Photo-Dissociation of Diatomic Molecules by Chirped Pulses

So far we have considered the excitation of energy levels. In the case of real chemical dynamics, however, we have to deal with multi-dimensional problems, namely, pump or dump of wave packet between two potential energy surfaces has to be dealt with. The idea of periodic chirping discussed so far is applicable to these cases, although the efficiency cannot be 100% anymore. Since the wave packet moves on the potential energy surface, the chirping should be quickly finished before the wave packet moves away from the transition region. This is discussed here.

In order to control chemical dynamics, it is crucial to control wave packet motions. Here let us consider efficient electronic excitation of wave packet by ultrashort broadband laser pulses. Although it is possible to use two linearly chirped pulses [9], it is more efficient to use one quadratically chirped pulse [3, 4, 13, 14]. We assume that the nuclear configuration does not change during the electronic transition and regard the system as a coordinate-dependent energy level problem with the effect of the kinetic energy operator taken into account as a perturbation. Then the laser parameters can be designed again by using the Zhu-Nakamura theory (Chapter 4).

The total excitation probability from the ground state can be approximately expressed as

$$\mathcal{P} = \int P_{12}(x)|\Psi_g(x, t = 0)|^2 dx, \tag{5.66}$$

where $\Psi_g(x, t = 0)$ is the initial wave packet on the ground state and the nonadiabatic transition probability $P(x)$ is calculated from the corresponding two-level problem that depends on x parametrically. By taking into account the kinetic energy operator as a perturbation, the Floquet Hamiltonian is given by [3, 4, 13, 14]

$$H^F = \frac{1}{2}\begin{pmatrix} \hbar\omega(t) - \tilde{\Delta}(x) & -\mu\epsilon(t) \\ -\mu\epsilon(t) & -\hbar\omega(t) + \tilde{\Delta}(x) \end{pmatrix}, \tag{5.67}$$

where

$$\tilde{\Delta}(x) = \Delta(x) + \Delta t\vec{v} \cdot \nabla\Delta(x) \tag{5.68}$$

with

$$\Delta(x) = V_e(x) - V_g(x). \tag{5.69}$$

Here $V_j(x)(j = e, g)$ are the ground and excited potentials, $\vec{v}$ is the mean velocity of the wave packet, μ is the transition dipole moment, ϵ is the laser pulse envelope, and Δt is the time delay measured from the pulse center. The nonadiabatic transition probability $P(x)$ is expressed as usual from the Zhu-Nakamura theory as (see Chapter 4)

$$P(x) = 4p_{ZN}(1 - p_{ZN}) \sin^2 \psi_{ZN}. \tag{5.70}$$

The two basic parameters $\alpha \equiv a^2$ and $\beta \equiv b^2$ in the Zhu-Nakamura formulas in Chapter 4 are defined as

$$\alpha = \frac{\hbar\alpha_\omega}{(\mu\epsilon)^3} \tag{5.71}$$

$$\beta = \frac{\Delta(x) - \hbar\beta_\omega + \frac{(\vec{v}\cdot\nabla\Delta(x))^2}{4\hbar\alpha_\omega}}{\mu\epsilon}. \tag{5.72}$$

The parameters α_ω and β_ω are the chirping rate and the carrier frequency, namely, the laser frequency is chirped quadratically as

$$\omega(t) = \alpha_\omega(t - t_p)^2 + \beta_\omega. \tag{5.73}$$

The laser parameters should be chosen so that the probability $\mathcal{P}$ becomes as close to unity as possible. Fig. 5.14 depicts the probability P_{12} as a function of α and β. There are some areas where the probability is larger than 0.9, such as those around ($\alpha = 1.20$, $\beta = 0.85$), ($\alpha = 0.53, \beta = 2.40$), ($\alpha = 0.38, \beta = 3.31$) and so on. Since the potential energy difference $\Delta(x)$ and the transition dipole

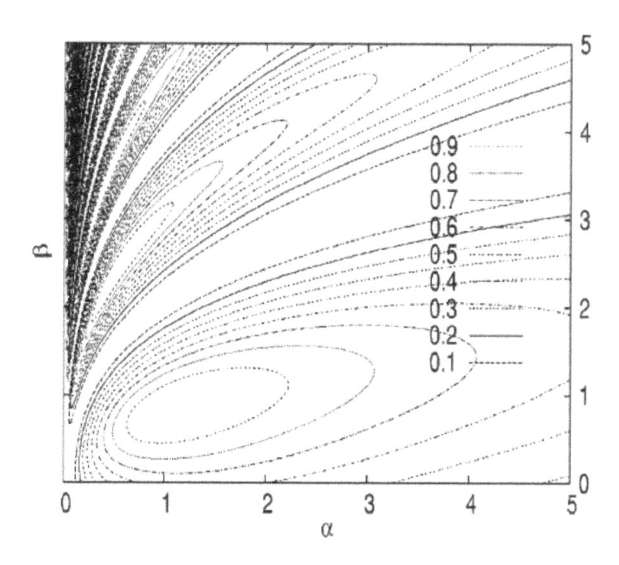

Figure 5.14: Contour map of the nonadiabatic transition probability P_{12} induced by a quadratically chirped pulse as a function of the two basic parameters α and β. Reproduced with permission from [13, 14].

moment $\mu(x)$ are coordinate-dependent, it is generally impossible to achieve perfect excitation by a single quadratically chirped pulse. A high efficiency of the population transfer is, however, possible without significant deformation of the shape of the wave packet, if we locate the parameters inside one of these islands. The biggest, and thus the most useful, island is around $(\alpha = 1.20, \beta = 0.85)$. The probability P_{12} is larger than 0.9, if $\alpha \in (0.62, 2.21)$ and $\beta \in (0.45, 1.30)$. Thus the condition for nearly complete excitation is given by

$$0.70 \lesssim \alpha \lesssim 2.0 \tag{5.74}$$

and

$$0.50 \lesssim \beta \lesssim 1.20. \tag{5.75}$$

In the following sections two numerical examples are shown below. The envelope of laser pulse commonly employed is the tangent hyperbolic type,

$$\epsilon(t) = \frac{\epsilon_0}{2}\left(\tanh\left[\frac{t - t_c + \tau/2}{s}\right] - \tanh\left[\frac{t - t_c - \tau/2}{s}\right]\right), \tag{5.76}$$

where t_c, s, τ, and ϵ_0 are the center time, switching time, duration, and maximum amplitude, respectively. The center time is taken to be equal to the frequency center time, that is, $t_c = t_p$ (see Eq. (5.73)).

5.3.1 *Photo-Dissociation of* LiH *from a Non-equilibrium Displaced Position*

A displaced wave packet on the ground state $X^1\Sigma^+$ at $R = 6.0$ *a.u.* is excited to the $B^1\Pi$ excited state (see Fig. 5.15) [3, 4, 13]. This displaced wave packet is used to demonstrate that the method of quadratically chirped pulse works even for a temporary unstable initial state. The potential energy curves and the transition dipole moment are taken from [15]. This wave packet is not simply put at $R = 6.0a_0$ at $t = 0$, but is prepared by moving the ground vibrational state by applying a sequence of quadratically chirped IR pulses (see Fig. 5.16). Thus this demonstrates that we can move a wave packet on the same electronic potential surface. The time dependence of the wave packet population on the X and B states is

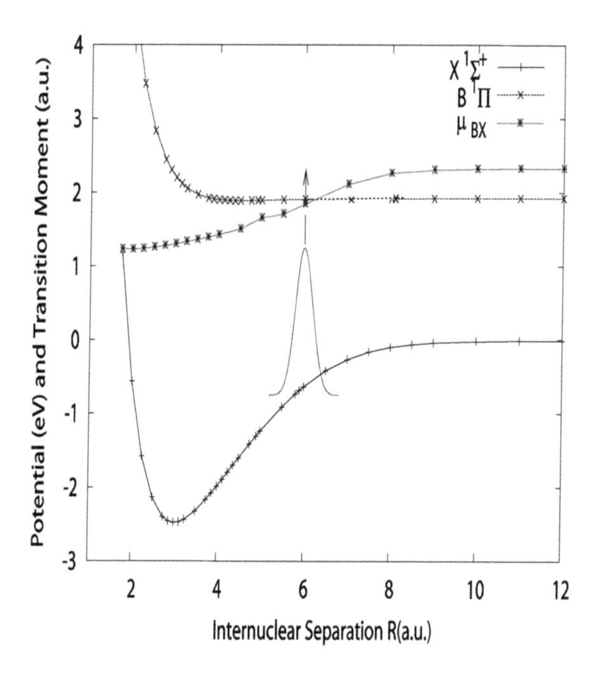

Figure 5.15: Pumping of a LiH wave packet from the outer classical turning point ($\sim 6a_0$) of the ground $X^1\Sigma^+$ state. The wave packet is the one shifted from the ground vibrational state and the $X \rightarrow B$ transition is considered. Reproduced with permission from [13].

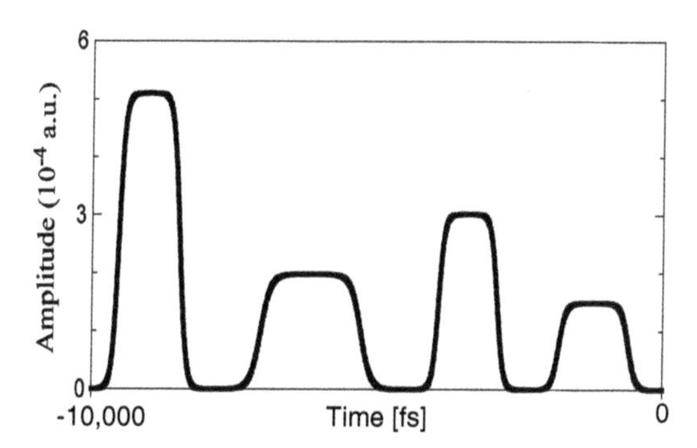

Figure 5.16: Envelope of the quadratically chirped pulses to move the ground vibrational state to the initial wave packet shown in Fig. 5.15.

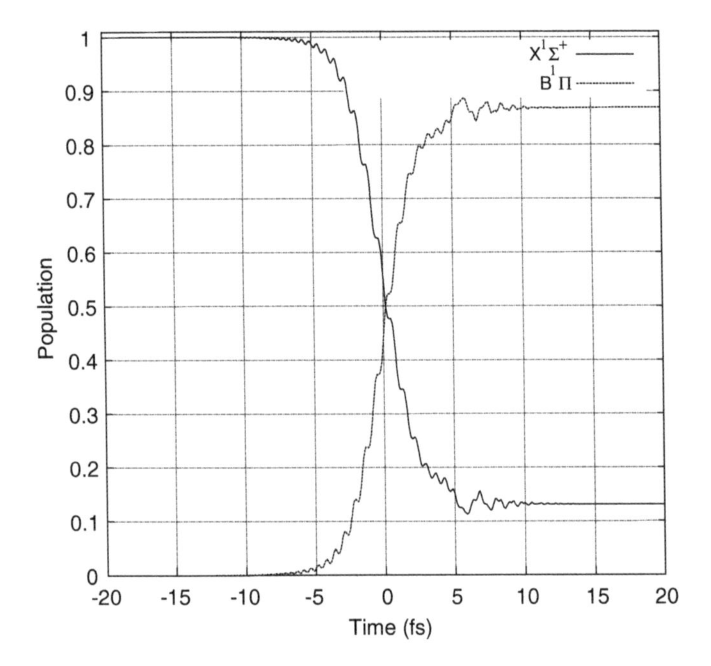

Figure 5.17: Time variation of the wave packet population on the ground X and excited B states of LiH. The laser is a quadratically chirped pulse with parameters: $\alpha_\omega = 5.84 \times 10^{-2}$ eVfs^{-2}, $\beta_\omega = 2.319$ eV, and $I = 1.00$ TWcm^{-2}. The pulse is centered at $t = 0$ and has a temporal width $\tau = 20$ fs. Reproduced with permission from [13].

plotted in Fig. 5.17. The excitation takes place in a few femtoseconds and the integrated total transition probability given by Eq. (5.66) is $\mathcal{P} = 0.879$, which is in good agreement with the value 0.864 obtained by numerical solution of the original coupled Schrödinger equations. This means that the population deviation from 100% is not due to the approximation, but comes from the intrinsic reason, that is, from the spread of the wave packet. The excited wave packet on the B state moves out to the asymptotic region and a high dissociation efficiency is attained. Note that the LiH molecule is one of the most difficult systems to apply the present method to, since the mass of LiH is very light and the gradient of potential difference is relatively large. These difficulties can be overcome by using the quick quadratic chirping. The NaK molecule is heavy and presents another extreme case. This is discussed in the next section.

5.3.2 *Photo-Dissociation of* NaK

This represents an example of the quadratically chirped pump-dump scheme [3, 4, 13, 14]. Fig. 5.18 shows the potential curves of the ground state $X^1\Sigma^+$ and the excited state $A^1\Sigma^+$. The transition dipole moment is also shown. These are taken from [16]. The whole process is as follows:

(i) The ground vibrational state of the ground electronic state $X^1\Sigma^+$ is pumped up to the electronically excited state $A^1\Sigma^+$.
(ii) This wave packet on the excited state is dumped down to the ground state at the outer turning point.
(iii) When this dumped wave packet comes to the inner turning point of the ground electronic state, this is finally pumped up again to the electronically excited state, and
(iv) the wave packet dissociates along the excited state.

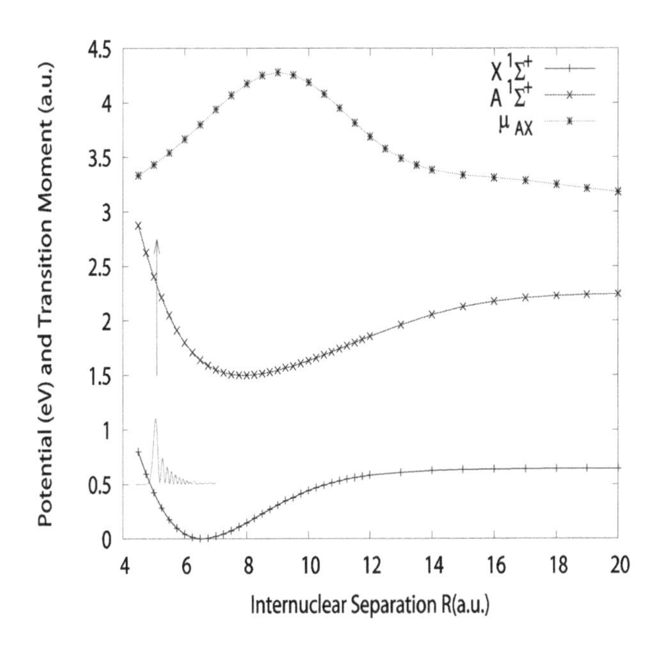

Figure 5.18: Electronic excitation of the NaK wave packet from the inner turning point on the ground electronic state X. The $X \rightarrow A$ transition is considered. The initial wave packet is prepared by two quadratically chirped pulses by the pump-dump mechanism (see the procedures (i)–(iii) in the text and Figs. 5.20 and 5.21). Reproduced with permission from [13].

The procedures (i) and (ii) are employed, simply because the direct excitation from the ground vibrational state cannot lead to dissociation on the excited electronic state A. The wave packet in Fig. 5.18 shows the one at the inner turning point of the electronic ground state X prepared by the above processes (i)–(iii). Although this initial wave packet is quite bumpy, the calculated total excitation probability $\mathcal{P}$ is as high as $\mathcal{P} = 0.905$ at the laser intensity of only 0.2 $\mathrm{TW/cm^{-2}}$ (see Fig. 5.19).

The quadratically chirped pump-dump scheme of the processes (i) and (ii) mentioned above is demonstrated next. Since the pioneering work by Tannor and Rice [17], the pump-dump method has been widely used to control various processes. However, since it is not possible to transfer a wave packet from one potential energy surface

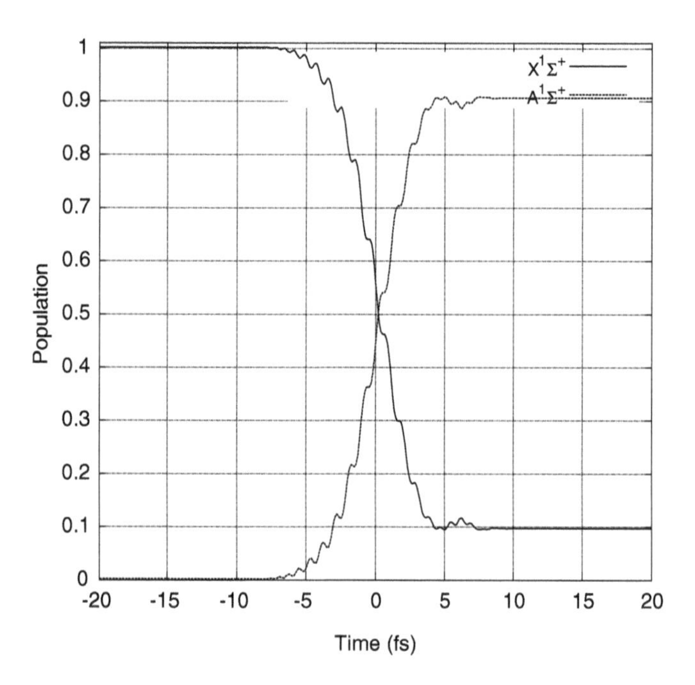

Figure 5.19: Time variation of the wave packet population on the ground X state and the excited A state of NaK. The laser is a quadratically chirped pulse with parameters: $\alpha_\omega = 3.13 \times 10^2$ eVfs^{-2}, $\beta_\omega = 1.76$ eV, and $I = 0.20$ TWcm^{-2}. The pulse is centered at $t = 0$ and has a temporal width $\tau = 20$ fs. Reproduced with permission from [13].

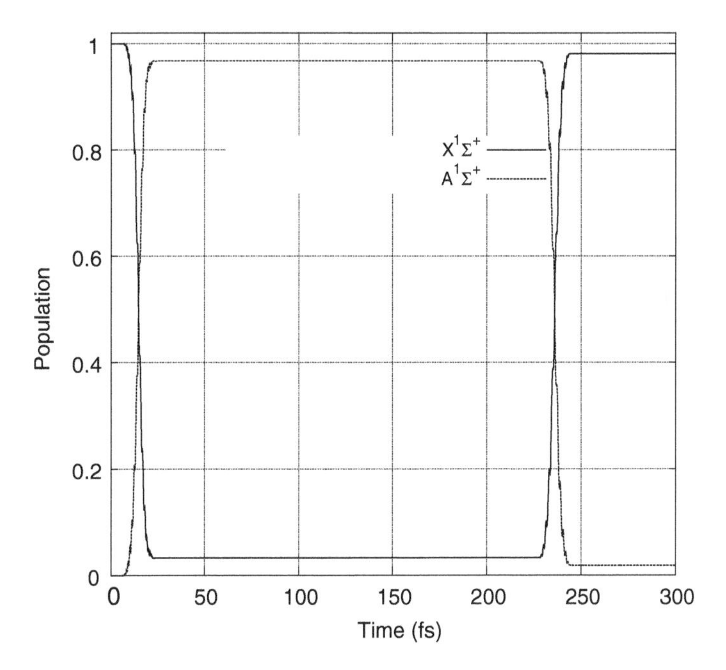

Figure 5.20: Pump-dump control of NaK molecule by using two quadratically chirped pulses. The initial wave packet of the ground vibrational eigenstate on the ground state X is pumped up to the excited state A. This excited wave packet is dumped at the outer turning point at $t \simeq 230$ fs by the second quadratically chirped pulse. The laser parameters are: $\alpha_\omega = 2.75(1.972) \times 10^{-2}$ eVfs^{-2}, $\beta_\omega = 1.441(1.031)$ eV, and $I = 0.15(0.10)$ TWcm^{-2} for the first (second) pulse. The two pulses are centered at $t_1 = 14.5$ fs and $t_2 = 235.8$ fs, respectively. Both of them have a temporal width $\tau = 20$ fs. Reproduced with permission from [13].

to another nearly completely by using the ordinary transform limited or linearly chirped pulses, the efficiency of the ordinary pump-dump method cannot be high. On the other hand, the present quadratic chirping method makes it possible to achieve nearly complete pump and dump. The time variation of the population by the processes (i) and (ii) is shown in Fig. 5.20. The overall pump-dump probability is found to be $\gtrsim$0.981. The final wave packet that arrived at the left turning point is nothing but the one shown in Fig. 5.18. This pump-dump method can be applied even to a wave packet moving in between the turning points, if its velocity is not too high. An example is shown in Fig. 5.21. The first step of pump is the same

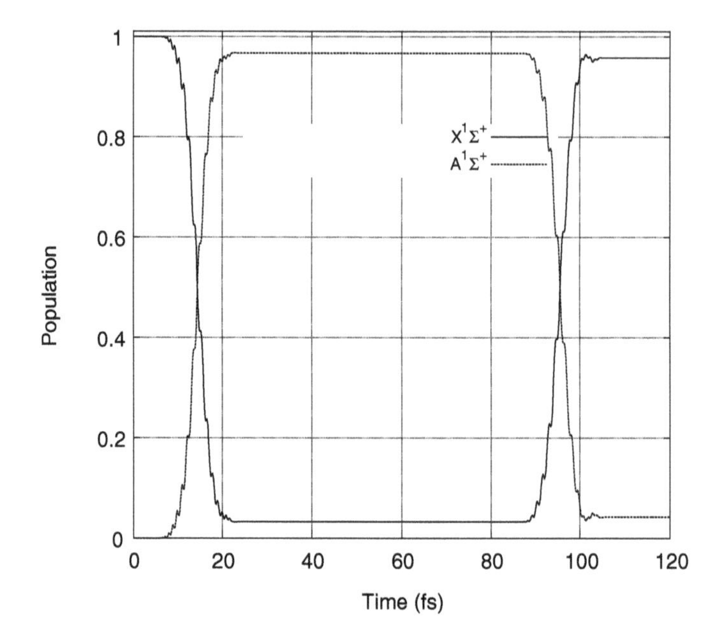

Figure 5.21: Pump-dump control of NaK molecule by using two quadratically chirped pulses. The initial state and the first step are the same as in Fig. 5.20. The excited wave packet is now dumped at $R \simeq 6.5\ a_0$ on the way to the outer turning point. The parameters of the second pulse are: $\alpha_\omega = 1.929 \times 10^{-2}$ eVfs^{-2}, $\beta_\omega =$ 1.224 eV, and $I = 0.10$ TWcm^{-2}. The second pulse is centered at $t_1 = 95.5$ fs and has a temporal width $\tau = 20$ fs. Reproduced with permission from [13].

as that in Fig. 5.20. This excited wave packet is now dumped down at $R \simeq 6.5a_0$ on the way to the outer turning point. The overall pump-dump probability is as high as 0.958.

References

[1] Y. Teranishi and H. Nakamura, *Phys. Rev. Lett.* **81**, 2032 (1998).
[2] Y. Teranishi and H. Nakamura, *J. Chem. Phys.* **111**, 1415 (1999).
[3] H. Nakamura, *Nonadiabatic Transition: Concepts, Basic Theories and Applications*, World Scientific, 2012 (2nd edition).
[4] H. Nakamura, *Introduction to Nonadiabatic Dynamics*, World Scientific, 2019.
[5] Y. Teranisi and H. Nakamura, *J. Chem. Phys.* **107**, 1904 (1997).
[6] E.E. Nikitin and S.Ya. Umanskii, *Theory of Slow Atomic Collisions*, Springer, 1984.

[7] V.I. Osherov and H. Nakamura, *J. Chem. Phys.* **105**, 2770 (1996).
[8] V.I. Osherov, V.G. Ushakov and H. Nakamura, *Phys. Rev.* **A57**, 2672 (1998).
[9] K. Nagaya, Y. Teranishi and H. Nakamura, *J. Chem. Phys.* **117**, 9588 (2002).
[10] K. Nagaya, Y. Teranishi and H. Nakamura, Chapter 7 in *Laser Control and Manipulation of Molecules*, American Chemical Society, 2002, pp. 98–117.
[11] M. Sugawara and Y. Fujimura, *J. Chem. Phys.* **100**, 5646 (1994).
[12] A. Kondorskiy, S. Nanbu, Y. Teranishi and H. Nakamura, *J. Phys. Chem.* **114**, 6171 (2010).
[13] S. Zou, A. Kondorskiy, G. Mil'nikov and H. Nakamura, *J. Chem. Phys.* **122**, 084112 (2005).
[14] S. Zou, A. Kondorskiy, G. Mil'nikov and H. Nakamura, Chapter 5 in *Progress in Ultrafast Intense Laser Science II*, Springer, 2007, pp. 95–117.
[15] H. Partridge and S.R. Langhoft, *J. Chem. Phys.* **74**, 2361 (1981).
[16] S. Magnier, M.A. Aubert-Frecon and Ph. Millie, *J. Mol. Spec.* **200**, 86 (2000).
[17] D.J. Tannor and A. Rice, *J. Chem. Phys.* **83**, 5013 (1985).

Chapter 6

Complete Reflection Phenomenon

6.1 Basic Idea

The transition in the nonadiabatic tunneling type of potential system (see Fig. 4.2) presents quite a unique and intriguing phenomenon. This potential system creates a potential barrier which cannot be treated in the same way as in the ordinary potential barrier even at energies lower than the barrier top and the phenomenon of complete reflection occurs at energies higher than the bottom of the upper adiabatic potential. The nonadiabatic tunneling probability is always smaller than the transmission probability through the corresponding single potential barrier with the upper adiabatic potential neglected (see Fig. 1.1). This means that the existence of the upper adiabatic potential cannot be neglected even at energies lower than the top of the lower adiabatic potential. Furthermore, the nonadiabatic tunneling probability oscillates as a function of energy and at certain discrete energies the transmission probability becomes exactly zero. This is called complete reflection. The transmission probability $P = |T|^2$ at energies higher than the bottom of the upper adiabatic potential, i.e., $E \geq E_b$, is given by (see Chapter 4)

$$P = |T|^2 = \frac{4\cos^2 \psi_{\mathrm{ZN}}}{4\cos^2 \psi_{\mathrm{ZN}} + p_{\mathrm{ZN}}^2/(1 - p_{\mathrm{ZN}})}, \tag{6.1}$$

where T is the transmission amplitude, and p_{ZN} is the nonadiabatic transition probability of the Zhu-Nakamura theory for one passage of the crossing point. This equation clearly shows that complete reflection ($P = 0$) occurs when the following condition is satisfied:

$$\psi_{\rm ZN} = \left(l + \frac{1}{2}\right)\pi \quad \text{for } l = 0, 1, 2, \ldots. \tag{6.2}$$

The transmission amplitude T is nothing but the reduced scattering matrix $(S^{\rm R}_{NT})_{12}$ that is explicitly expressed as

$$T = (S^{\rm R}_{NT})_{12} = \frac{2\sqrt{1 - p_{\rm ZN}}e^{i\sigma_{\rm ZN}}}{1 + (1 - p_{\rm ZN})e^{2i\psi_{\rm ZN}}} \cos\psi_{\rm ZN}, \tag{6.3}$$

which can be easily reduced to Eq. (6.1). Equation (6.3) can be easily rewritten as

$$\begin{aligned} T &= \sqrt{1 - p_{\rm ZN}}e^{i\phi_{\rm S}}\left(1 - \frac{p_{\rm ZN}}{p_{\rm ZN} - 1 - e^{-2i\psi_{\rm ZN}}}\right) \\ &= (I_X)_{11} + \frac{(I_X)_{12}L_{\rm r}^2(O_X)_{22}L_{\rm l}^2(I_X)_{21}}{1 - (O_X)_{22}L_{\rm l}^2(I_X)_{22}L_{\rm r}^2}, \end{aligned} \tag{6.4}$$

where $(I_X)_{nm}$ is the (n, m)-element of the matrix I_X, representing the nonadiabatic transition amplitude from the adiabatic state m to the state n at the avoided crossing point when the wave propagation proceeds inward, namely from the right to the left. The matrix O_X is the transpose of I_X, and represents the transition in the outgoing (left to right) segment. $L_{\rm l}$ and $L_{\rm r}$ are defined by (see Fig. 4.2)

$$L_{\rm l} = \exp\left[i\int_{T_2^l}^{R_X} K_2(x)dx + i\frac{\pi}{4}\right] \tag{6.5}$$

and

$$L_{\rm r} = \exp\left[i\int_{R_X}^{T_2^r} K_2(x)dx + i\frac{\pi}{4}\right]. \tag{6.6}$$

This equation and the first equation in Eq. (6.4) enable us to clarify the physical meaning of this complete reflection phenomenon. The first term $(I_X)_{11}$ represents the wave which simply transmits the barrier without any transition to the upper adiabatic state $E_2(x)$, and the second term corresponds to the transmitting wave which is trapped by the upper adiabatic potential. At the energies of complete reflection $[\exp(-2i\psi_{ZN}) = -1]$ the second term in the

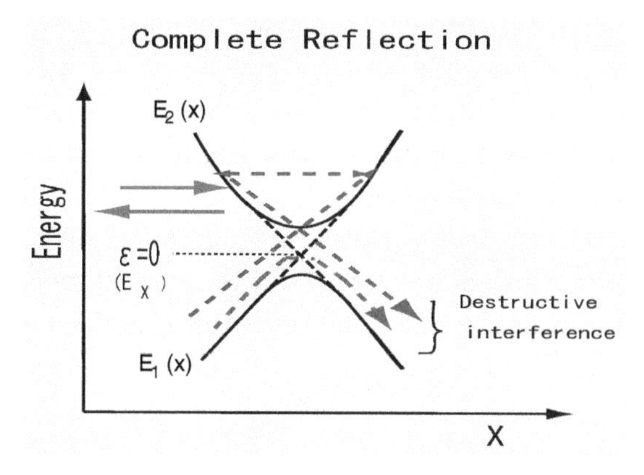

Figure 6.1: Physical interpretation of the phenomenon of complete reflection.

bracket of the first equation becomes exactly unity and cancels the first term whatever the probability p_{ZN} is. This means that these waves interfere destructively at the exit whatever the nonadiabatic transition probability $p_{\rm ZN}$ is and the incident wave is completely reflected back (see Fig. 6.1). The complete reflection condition given by Eq. (6.2) is similar to the famous Bohr–Sommerfeld quantization condition. In the present case, however, the effect of nonadiabatic coupling naturally appears as the additional phase $\phi_{\rm S}$, which tends to $\pi/4$ (0) in the weak diabatic coupling limit $\delta_{\rm ZN} \to 0$ (in the strong diabatic coupling limit $\delta_{\rm ZN} \to \infty$).

This unique phenomenon of complete reflection suggests intriguing possibilities such as bound states in the continuum and molecular switching in a periodic potential system [1, 2].

In the following, in order to explain how to use the complete reflection phenomenon to control dynamics, a one-dimensional model of photo-dissociation is discussed here by taking a linear triatomic molecule ABC consisting of a bound ground electronic state and a dissociative excited electronic state with two dissociation channels, A+BC and AB+C [3]. Complete and selective dissociation into any desired channel can be realized by adjusting the initial vibrational excited state and the laser frequency. Dissociation can be realized even when a potential barrier hinders the ordinary photo-dissociation along the electronically excited state potential surface. This control scheme cannot be very robust, because the method is based on the phase interference, but this can present a new intriguing selective control of molecular photo-dissociation when the conditions with respect to potential energy surface topography and initial vibrational state are appropriately satisfied.

In order to demonstrate the present idea, numerical calculations of wave packet propagation are carried out for a one-dimensional potential system shown in Fig. 6.2. The potential functions employed are

$$V_1(r) = 0.2(1 - e^{-1.5r^2}) \tag{6.7}$$

and

$$V_2(r) = \begin{cases} 0.15e^{-2.0(r-0.25)^2} + 0.25 & (r > 0.25) \\ 0.18e^{-2.0(r-0.25)^2} + 0.22 & (r \leq 0.25) \end{cases} \tag{6.8}$$

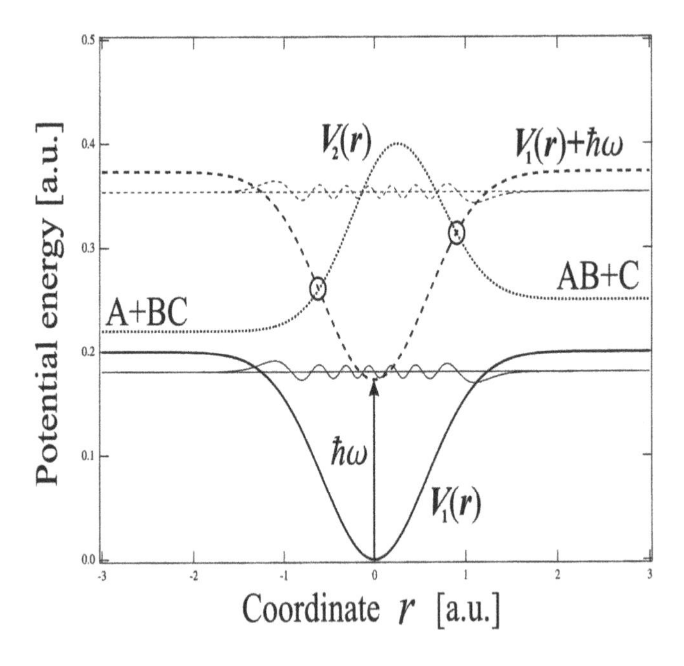

Figure 6.2: One-dimensional model of the control scheme. Solid line: ground electronic state $V_1(r)$, dotted line: excited electronic state $V_2(r)$. Two circles represent the nonadiabatic tunneling-type crossings created between the excited electronic state and the dressed ground electronic state (dashed line). The 14th vibrational eigenstate (thin line) and its dressed state (thin dashed line) of $V_1(r)$ are also depicted. Reproduced with permission from [3].

in atomic units. The molecule-laser interaction is taken to be

$$V_{\text{int}}(t) = -\mu E(t) = \mu\sqrt{I}\cos(\omega t + \delta), \tag{6.9}$$

where μ is the transition dipole moment between the two electronic states and $E(t)$ is the stationary laser field with the frequency ω and the intensity I.

The 14th vibrational eigenstate of the ground electronic state (the quantum number $v = 13$) is prepared as an initial state and the following parameters are used for the wave packet calculation: m (reduced mass) $= 1.0$ a.m.u., $\mu = 1.0$ a.u., $I = 1.0$ TW/cm^2, Δt (time step) $= 1.0$ a.u., and Δr (spatial grid size) $= 0.016$ a.u. The total number of the spatial grid points is 512. The complete reflection positions can be analytically predicted as shown in Fig. 6.3. Once the laser intensity I is selected, the shapes of the adiabatic potential curves, i.e., the ground electronic state shifted up by one

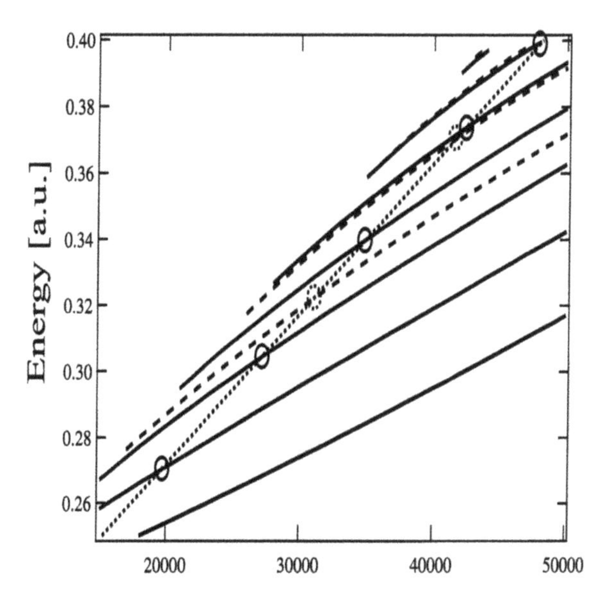

Figure 6.3: Analytical prediction of the complete reflection positions. Solid (dashed) lines: complete reflection positions at the left-side (right-side) channel. Dotted line: dressed 14th vibrational eigen state of $V_1(r)$. Solid (dashed) circles represent the complete reflection of the vibrational state at the left-side (right-side) crossing. Reproduced with permission from [3].

photon energy $V_1(r) + \hbar\omega$ and the excited electronic state $V_2(r)$, are determined for a fixed laser frequency ω and then the positions of complete reflection can be easily estimated from Eq. (6.2) (see also Eq. (6.1)). The solid (dashed) line in Fig. 6.3 represents the complete reflection position at the left-side (right-side) channel. The dotted line represents the relevant vibrational level, which is shifted up by one photon energy $\hbar\omega$, namely a linear function of the frequency ω. Thus the crossing points marked by solid (dotted) circles give the positions of the complete reflection on the left (right) side for this vibrational state. The calculated dissociation probabilities against laser frequency are depicted in Fig. 6.4. In order to concentrate on the dissociation dynamics, here we have assumed that the initial vibrational state was prepared. The solid (dotted) line represents the dissociation into the right (left) channel; the zero probability positions of the solid (dotted) line coincide with the solid (dotted)

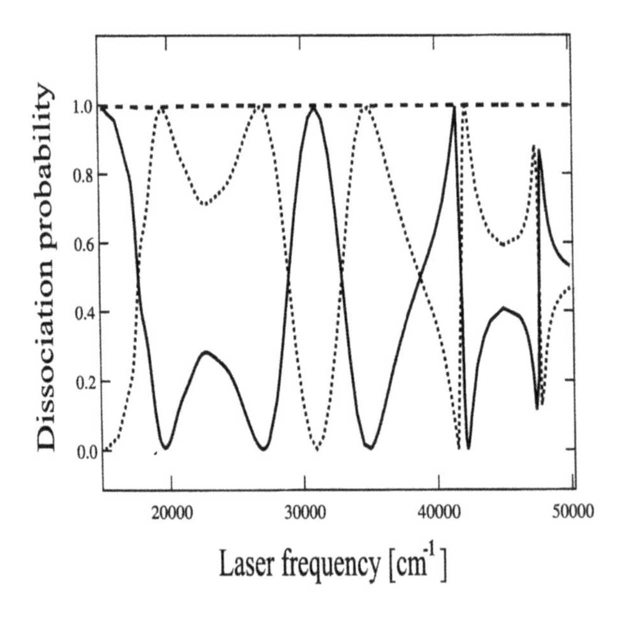

Figure 6.4: Dissociation probability as a function of laser frequency. Solid line: dissociation into the right-side (AB+C) channel, dotted line: dissociation into the left-side (AB+C) channel, dashed line: sum of the two dissociation probabilities to guarantee the unitarity. Reproduced with permission from [3].

circles in Fig. 6.3. The highest dip at $\omega \sim 48000$ cm^{-1} is not complete because of tunneling. As is demonstrated above, within the one-dimensional model the present control scheme is perfect and a molecule can be dissociated into any channel we desire by adjusting the laser frequency for a given vibrational state. The initial vibrational state should, however, be an excited one, since the phase ψ should satisfy Eq. (6.2).

6.2 Photo-Dissociation Branching of HI

The selective photo-dissociation of an HI molecule with use of the complete reflection phenomenon is considered in the energy range $\hbar\omega = 3$–6 eV [4]. In the case of diatomic molecule, pre-dissociation can be stopped, if the following condition is satisfied (see Eqs. (6.1)–(6.2)):

$$\psi_{ZN}(E) = (n + 1/2)\pi \quad (n = 0, 1, 2, \ldots). \tag{6.10}$$

In this energy range, there are three electronically excited states $^1\Pi_1, ^3\Pi_{0+}$ and $^3\Pi_1$, as shown in Fig. 6.5. The ground state $^1\Sigma$ is coupled to these excited states by the transition dipole moments (see Fig. 6.6). Since the coupling among the excited states can be neglected because of the off-resonance condition, we have the following potential matrix:

$$V(R,t) = \begin{pmatrix} V_1(R) & -\mu_{12}(R)E(t) & -\mu_{13}(R)E(t) & -\mu_{14}(R)E(t) \\ -\mu_{12}(R)E(t) & V_2(R) & 0 & 0 \\ -\mu_{13}(R)E(t) & 0 & V_4(R) & 0 \\ -\mu_{14}(R)E(t) & 0 & 0 & V_4(R) \end{pmatrix}, \tag{6.11}$$

where V_j ($i = 1, 2, 3, 4$) correspond to $^1\Sigma, ^1\Pi_1, ^3\Pi_{0+}, ^3\Pi_1$, respectively. The potentials and dipole moments are taken from the *ab initio* data and spline fitting is applied to them [5, 6]. The CW laser field is taken as

$$E(t) = E_0(t)\cos(\omega t), \tag{6.12}$$

where $E_0(t)$ is the envelope of laser field that should be wide and smooth enough so that unnecessary transitions are not induced due

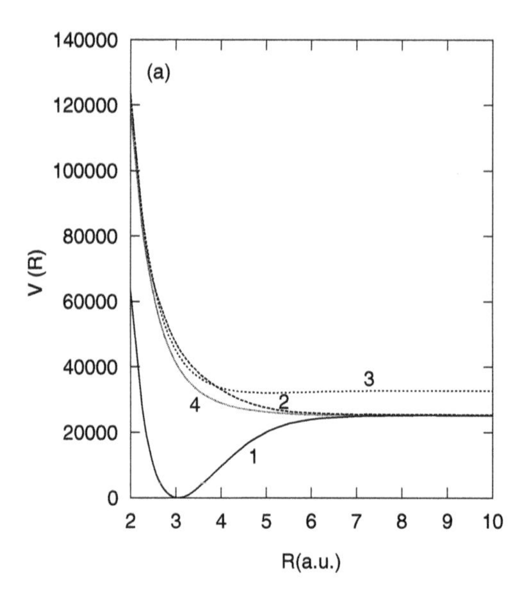

Figure 6.5: *Ab initio* potential energy curves of HI. The unit of $y-$axis is cm^{-1}. The potential curves $V_1(R)[^1\Sigma_+], V_2(R)[^1\Pi_1], V_3(R)[^3\Pi_{0+}]$ and $V_4(R)[^3\Pi_1]$ are designated as 1, 2, 3 and 4, respectively. Reproduced with permission from [4].

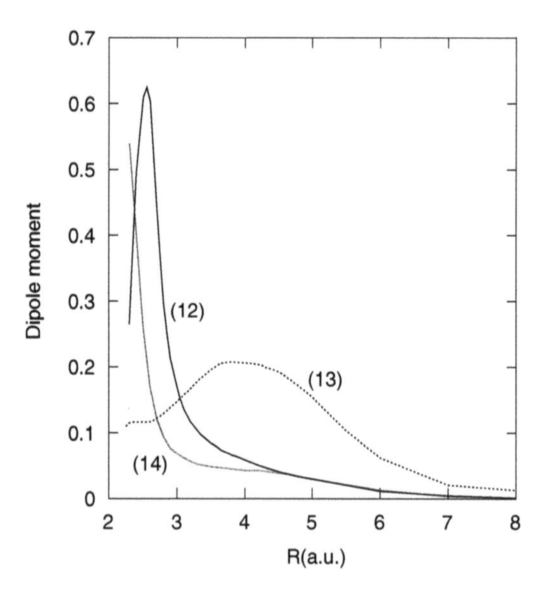

Figure 6.6: *Ab initio* transition dipole moments between the electronic ground and excited states of HI. The dipole moments $\mu_{1j}(j = 2, 3, 4)$ are designated as $(1j)$. Reproduced with permission from [4].

to sudden switching of the field. The actual intensity used in the calculations is 1.0 TW/cm^2. The two excited states $V_2 = {}^1\Pi_1$ and $V_4 = {}^3\Pi_1$ correlate to the ground state iodine I and the state $V_3 = {}^3\Pi_{0+}$ correlates to the excited iodine I*. Thus, in order to selectively produce the excited iodine I*, one has to be able to stop the dissociation through the two excited states V_2 and V_4. Numerical solutions are made by solving the time-dependent coupled equations,

$$i\hbar\frac{\partial}{\partial t}\phi(t) = \left(-\frac{\hbar^2}{2m}\frac{d^2}{dR^2} + V(R,t)\right)\phi(t). \tag{6.13}$$

The step sizes used are $\Delta R = 7.8 \times 10^{-3}$ a.u. and $\Delta t = 0.043$ fs, and the absorbing potential is put at $R = 9$–10 a.u. First, the complete reflection manifold in which Eq. (6.10) is satisfied is found and a rough estimate of appropriate energy region and vibrational states is made. One example of the results is shown in Figs. 6.7 and 6.8 for $v = 4$. Fig. 6.7 shows the time-integrated dissociation flux

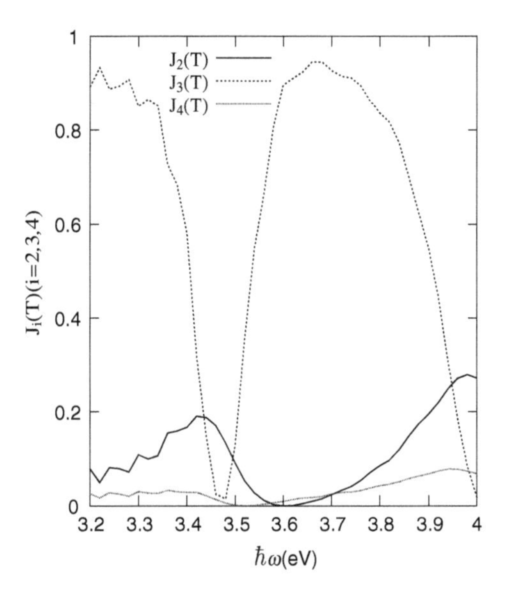

Figure 6.7: The time-integrated fluxes at $t = 3.5$ ps as a function of the photon energy for $v = 4$. The suffices $i = 2, 3, 4$ correspond to the excited states ${}^1\Pi_1, {}^3\Pi_{0+}, {}^3\Pi_1$, respectively. The fluxes J_2 and J_4 are almost stopped at $\hbar\omega \simeq 3.58$ eV. Reproduced with permission from [4].

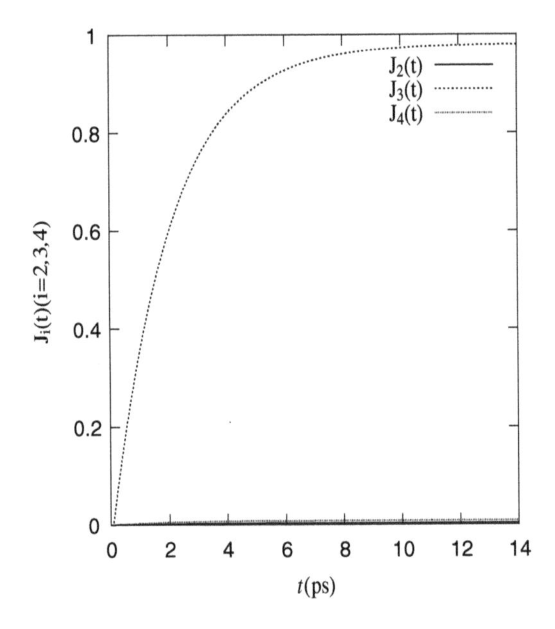

Figure 6.8: Time variation of the time-integrated fluxes $J_i(t)$ for $v = 4$. The suffices $i = 2, 3, 4$ correspond to the excited states ${}^1\Pi_1, {}^3\Pi_{0+}, {}^3\Pi_1$, respectively. The fluxes J_2 and J_4 are almost zero. Reproduced with permission from [4].

at $t = 3.5$ ps. The time-integrated flux is defined as

$$J_i(t) = \int_0^t dt \frac{\hbar}{m} \mathrm{Im} \left[\phi_i^*(R,t) \frac{d}{dt} \phi_i(R,t) \right]_{R=6}, \qquad (6.14)$$

where $\phi_i(R,t)$ is the wave packet on the potential i. It can be seen that the two dissociation channels are through the states $i = 2$ and 4 are almost stopped at the photon energy $\hbar\omega \simeq 3.58$ eV. Figure 6.8 shows the time-integrated flux to confirm the highly selective production of I*. The condition for producing I selectively can also be found easily. For example, $\hbar\omega \simeq 3.47$ eV with $v = 4$ meets the condition. In order to compare with any experiment, it is necessary to take into account the effects of initial rotational state distribution that depend on experimental condition. The completeness would be deteriorated to some extent, but the control may be achieved to a good extent. One defect of the method based on the complete reflection phenomenon is that the initial state should be prepared in a certain excited vibrational state.

6.3 Photo-Dissociation of CH_3SH

In this molecule, the saddle point of the excited electronic state is located in the $CH_3 + SH$ channel and blocks C–S bond breaking [7]. C–S bond breaking leading to the $CH_3 + SH$ product has been experimentally observed at wavelengths shorter than the first absorption band which leads to S–H bond breaking, but it remains as the minor pathway even though C–S cleavage is energetically favored [8]. Here this molecule is taken as another example to demonstrate the possibility of complete reflection phenomenon in two dimensions [3]. The ground state ($\tilde{X}^1A'$) and the first excited state ($1^1A''$) potentials are taken from [7]. With use of these potential energy surfaces, the two-dimensional wave packet calculations are performed. The initial vibrational state used is the 123rd one which corresponds to the $v = 8$ local mode of the S–H bond. In the same way as before, Fig. 6.9(a) presents the analytical prediction of the complete reflection positions in the $CH_3S + H$ channel based on the one-dimensional cut of the potential energy surface along the minimum energy path of the S–H bond and the vibrational state $v = 8$ of that bond. Figure 6.9(b) shows the dissociation flux accumulated over the time interval of 1 ps for S–H bond breaking, i.e., the dissociation into the CH_3S+H channel, as a function of the laser frequency. There are some conspicuous dips which correspond to the circles in Fig. 6.9(a). These dips are, of course, due to the complete reflection in the $CH_3S + H$ channel and indicate that the reflected wave packet can be transfered into the other channel $CH_3 + SH$ via the ground electronic state through mode-coupling. In the present model potential, however, there is no mode-coupling between the two channels in the ground electronic state; thus the wave packet stays inside and does not dissociate into the $CH_3 + SH$ channel. If there were an appropriate coupling between the two channels, we could dissociate this molecule into the unusual channel of $CH_3 + SH$ by using the present control scheme.

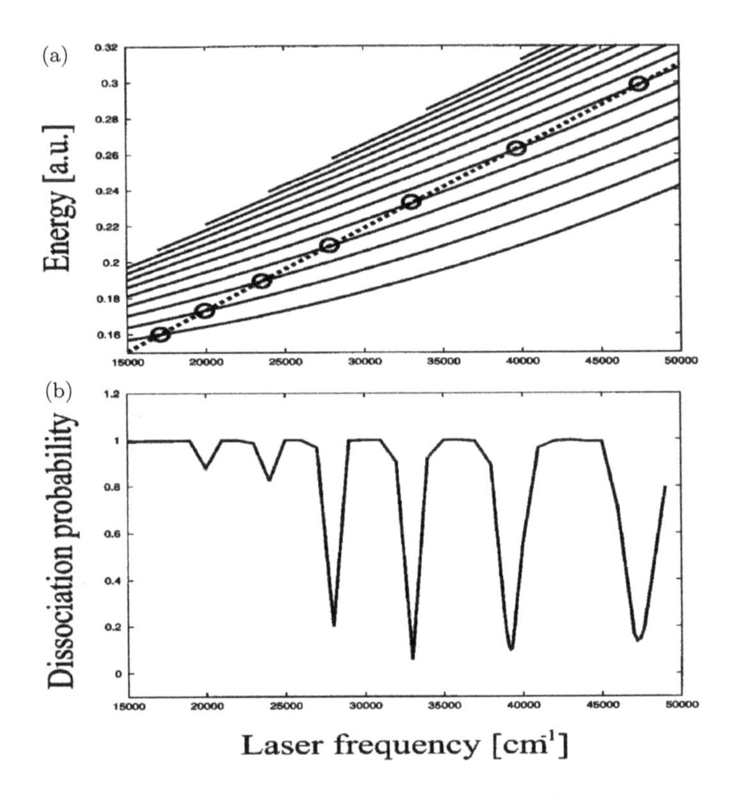

Figure 6.9: (a) Analytical prediction of the complete reflection positions in the case of a two-dimensional model of CH_3SH. Solid lines indicate the complete reflection positions in the CH_3S+H channel. Dashed line indicates the vibrational state $v = 8$ of the S–H bond. The solid circles represent the complete reflection positions for this 123rd vibrational eigenstate. (b) Dissociation probability against the laser frequency for $CH_3SH \rightarrow CH_3S + H$ in the case of the above vibrational state. Reproduced with permission from [3].

6.4 Photo-Dissociation of HOD

Let us next consider a two-dimensional model. As an example, we take the HOD molecule with two dissociation channels: H+OD and HO+D [3]. We consider the ground electronic state $\tilde{X}$ and the excited electronic state $\tilde{A}$. The bending and rotational motions are neglected for simplicity with the bending angle fixed at the equilibrium structure of the ground electronic state, i.e., at 104.52°. The ground electronic state potential $V_1(r_{\rm H}, r_{\rm D})$ (shown by dotted

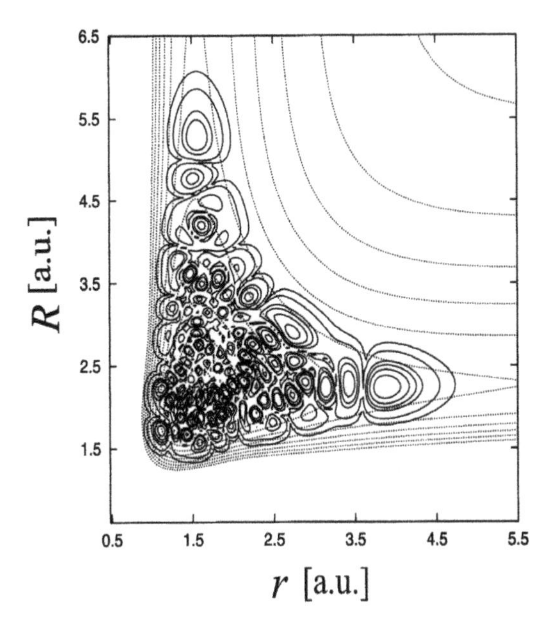

Figure 6.10: Contour plots of the ground electronic state of HOD (dotted line). The contour spacing is 8500 cm^{-1}. The density of the 145th vibrational eigenstate is superimposed (solid). Reproduced with permission from [3].

lines in Fig. 6.10) is taken to be two coupled Morse oscillators [9],

$$V_1(r_{\rm H}, r_{\rm D}) = D(1 - e^{-\gamma(r_{\rm H}-r_{\rm O})^2}) + D(1 - e^{-\gamma(r_{\rm D}-r_{\rm O})^2}) - B\frac{(r_{\rm H} - r_{\rm O})(r_{\rm D} - r_{\rm O})}{1 + e^{A((r_{\rm H}-r_{\rm O})+(r_{\rm D}-r_{\rm O}))}}, \quad (6.15)$$

where $r_{\rm H}$ ($r_{\rm D}$) represents the H–O (D–O) bond length, and $D = 0.2092$ hartree, $\gamma = 1.1327$ a.u.$^{-1}$, $r_0 = 1.81$ a.u., $A = 3.0$ a.u.$^{-1}$ and $B = 0.25$ hartree/a.u.^{2}. The last term represents the mode coupling, and the parameters A and B are assumed to be larger than the values used in [9] ($A = 1.0$ a.u.$^{-1}$ and $B = 0.00676$ hartree/a.u.^{2}) in order to demonstrate the present control scheme more clearly. The excited electronic state $V_2(r_{\rm H}, r_{\rm D})$ shown in Fig. 6.11 is taken to be the analytical function of Engel *et al.* [10], which was fitted to the *ab initio* calculations by Staemmler *et al.* [11]. The mass-scaled Jacobi coordinates r for O–H distance and R for OH–D distance are introduced,

$$r = \left(\frac{m_{\mathrm{OH}}}{m_{\mathrm{D,OH}}} \right)^{1/4} |\mathbf{r}_{\mathrm{H}}|, \tag{6.16}$$

$$R = \left(\frac{m_{\mathrm{D,OH}}}{m_{\mathrm{OH}}} \right)^{1/4} \left| \mathbf{r}_{\mathrm{D}} - \frac{m_{\mathrm{H}}}{m_{\mathrm{H}} + m_{\mathrm{O}}} \mathbf{r}_{\mathrm{H}} \right|, \tag{6.17}$$

where $\mathbf{r}_{\mathrm{H}}$ ($\mathbf{r}_{\mathrm{D}}$) is the vector from O atom to H (D) atom, $m_{\mathrm{OH}} = m_{\mathrm{O}}m_{\mathrm{H}}/(m_{\mathrm{O}}+m_{\mathrm{H}})$, $m_{\mathrm{D,OH}} = m_{\mathrm{D}}(m_{\mathrm{O}}+m_{\mathrm{H}})/(m_{\mathrm{D}}+m_{\mathrm{O}}+m_{\mathrm{H}})$, and m_{H}, m_{O} and m_{D} are the mass of H, O and D, respectively. With use of these coordinates (r, R) the two-dimensional time-dependent Schrödinger equation is written as

$$i\hbar \frac{\partial}{\partial t} \begin{bmatrix} \Psi_1(r,R,t) \\ \Psi_2(r,R,t) \end{bmatrix} = \begin{bmatrix} -\frac{\hbar^2}{2m}\left(\frac{\partial^2}{\partial r^2} + \frac{\partial^2}{\partial R^2}\right) + V_1(r,R) & -\mu E(t) \\ -\mu E(t) & -\frac{\hbar^2}{2m}\left(\frac{\partial^2}{\partial r^2} + \frac{\partial^2}{\partial R^2}\right) + V_2(r,R) \end{bmatrix} \times \begin{bmatrix} \Psi_1(r,R,t) \\ \Psi_2(r,R,t) \end{bmatrix}. \tag{6.18}$$

Here m is the reduced mass of the system,

$$m = \sqrt{\frac{m_{\mathrm{H}}m_{\mathrm{O}}m_{\mathrm{D}}}{m_{\mathrm{H}} + m_{\mathrm{O}} + m_{\mathrm{D}}}}. \tag{6.19}$$

$\Psi_1(r,R,t)(\Psi_2(r,R,t))$ represents the wave function of the ground (excited) electronic state. Equation (6.18) is solved by using the split operator method with the two-dimensional fast Fourier transform in the same way as before. The dissociation flux is integrated over time in a certain asymptotic region before the negative imaginary potentials which are put at both ends. The transition dipole moment μ is assumed to be 1.0 a.u. and the stationary laser field $E(t)$ is taken to be $\sqrt{I}\cos(\omega t + \delta)$. An initial state is prepared at the 145th vibrational eigenstate of $V_1(r,R)$ by solving the two-dimensional eigenvalue problem by the discrete variable

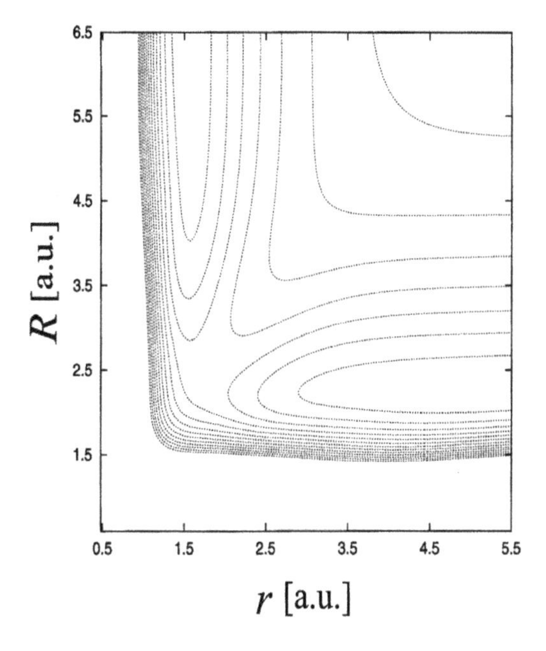

Figure 6.11: Contour plots of the excited electronic state of HOD. The contour spacing is 5000 cm^{-1}. Reproduced with permission from [3].

representation method [12]. This initial state is mainly composed of the 17th vibrational state of the O–H bond and the 23rd vibrational state of the O–D bond and spreads into both bonds due to the coupling, as is depicted by solid lines in Fig. 6.10. The laser intensity I and the various grid sizes are taken to be 1.0 TW/cm^2, $\Delta t = 1.0$ a.u., $\Delta r = 0.029$ a.u. and $\Delta R = 0.034$ a.u., respectively. The total number of spatial grid points is 256×256. Figure 6.12(a) depicts the calculated dissociation probabilities against laser frequency. The solid (dotted) line represents the dissociation probability into the H+OD (HO+D) channel. The dissociation probabilities change alternatively as a function of the laser frequency ω, and the dissociation into the H+OD (HO+D) channel is preferential when the laser frequency ω is $\sim$ 7000 cm^{-1}, 8000 cm^{-1}, 10000 cm^{-1}, and 11500 cm^{-1} (9000 cm^{-1}, 11000 cm^{-1} and 14000 cm^{-1}).

The control is not perfect this time because of multi-dimensionality, but is still quite selective. The molecule can be made to dissociate preferentially into any channel as we desire by choosing the laser frequency and the vibrational state appropriately. The complete reflection positions in the H+OD (HO+D) channel can be roughly estimated analytically by taking a one-dimensional cut of the potential energy surface along the minimum energy path of the O–H (O–D) bond into account and using the one-dimensional formula Eq. (6.2). The vibrational state is taken to be $v = 17(23)$ for the O–H (O–D) channel. Fig. 6.12(b) depicts these estimates. The solid (dashed) lines represent the complete reflection positions in the H+OD (HO+D) channel. The dotted (dash-dotted) line shows the vibrational state $v = 17$ of the O–H bond ($v = 23$ of the O–D bond) shifted up by one photon energy. Thus the solid (dotted) circles indicate the positions of complete reflection in the H+OD (HO+D) channel. As is seen in Fig. 6.12(a), the dips of calculated dissociation probabilities correspond well to the complete reflection positions predicted analytically. Exceptions are the dip at $\omega \sim 7200$ cm^{-1} and the peak at $\omega \sim 13500$ cm^{-1} in the H+OD dissociation channel. The former is shallow and shifted to higher frequency ($\sim$7500 cm^{-1}) in Fig. 6.12(a), and the latter has almost disappeared. This is due to the topography of the potential energy surface, representing the difficulty of multi-dimensionality.

The wave packet dynamics on the excited state are shown in Figs. 6.13 and 6.14. The wave packet is depicted by solid lines at various times. The dashed lines in these figures represent the crossing seam lines between the dressed ground state and the excited state. Fig. 6.13 corresponds to the laser frequency 9000 cm^{-1} and thus the wave packet moves out into the HO+D channel. Fig. 6.14 corresponds to the laser frequency 11500 cm^{-1}, and the wave packet almost dissociates into the H+OD channel (see Fig. 6.13).

As the above results demonstrate, selective dissociation based on the complete reflection phenomenon can be realized even in two-dimensional systems. The control naturally cannot be perfect like in the one-dimensional case, but can still be quite effective. The dissociation into a certain channel is stopped by the complete reflection phenomenon and the reflected wave packet is transferred into the

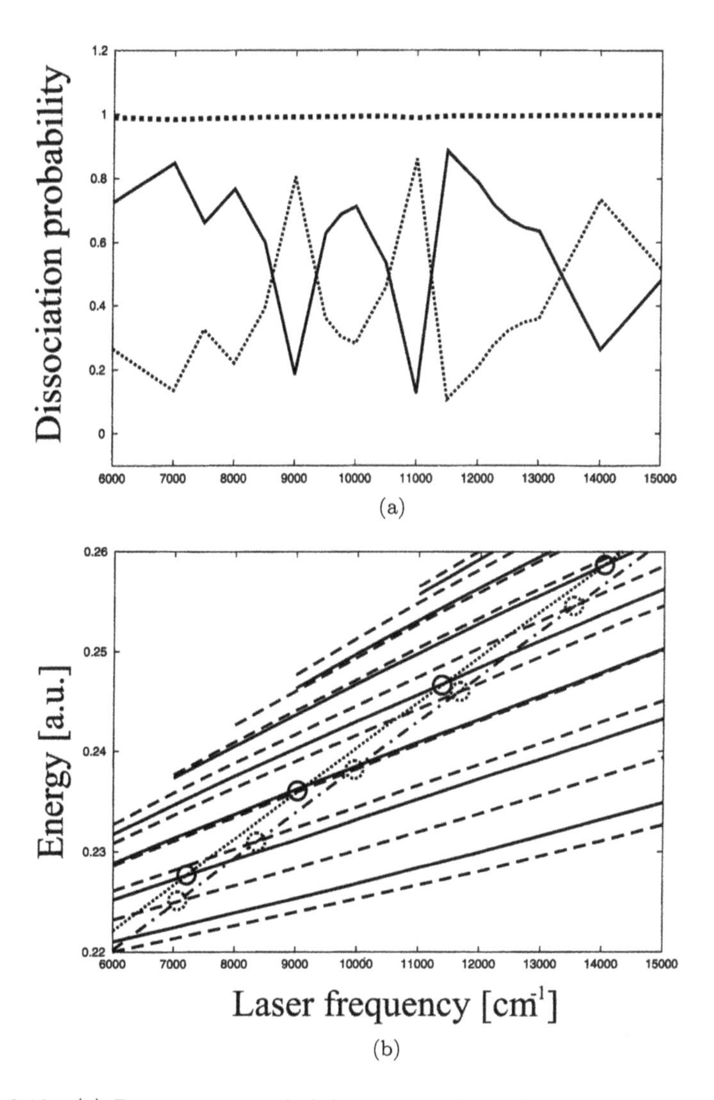

Figure 6.12: (a) Dissociation probability against laser frequency in the case of the 145th vibrational state of HOD. Solid (dotted) line: dissociation into the H+OD (HO+D) channel. Dashed line: sum of the two dissociation probabilities. (b) Analytical prediction of the complete reflection positions. Solid (dashed) lines: the complete reflection positions in the H+OD (HO+D) channel. Dotted (dash-dotted) line: the vibrational state $v = 17$ of the O–H bond ($v = 23$ of the O–D bond). The solid (dotted) circles represent the complete reflection positions when the 145th vibrational eigenstate is prepared as an initial state on the H+OD (HO+D) side. Reproduced with permission from [3].

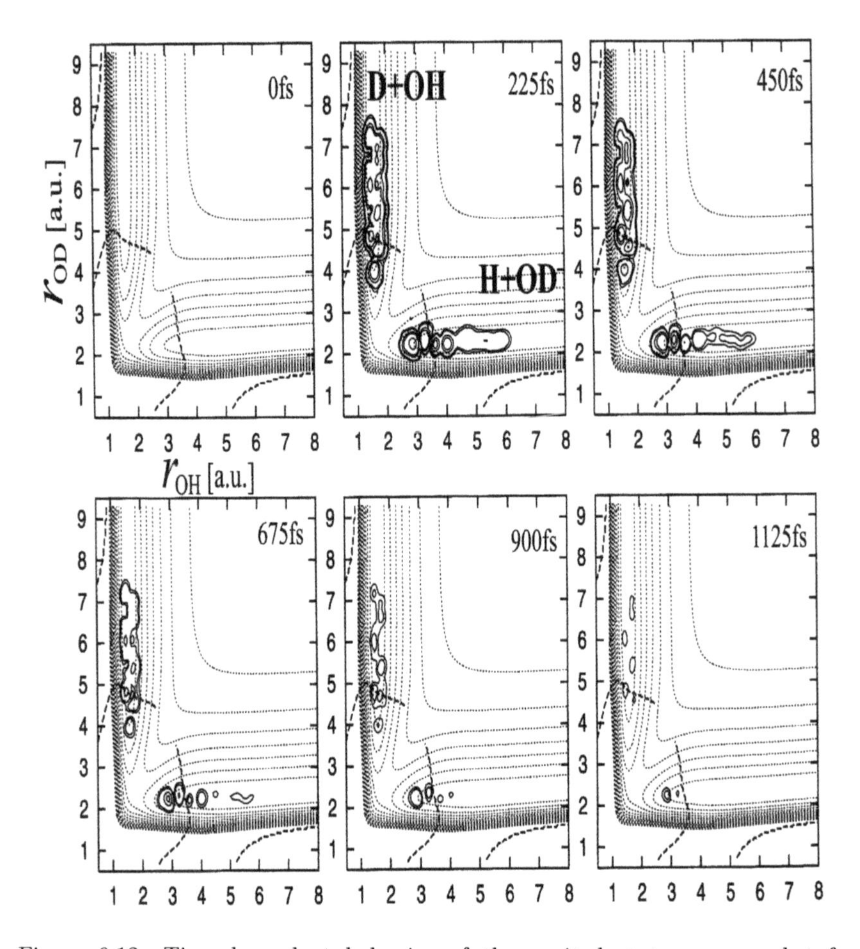

Figure 6.13: Time-dependent behavior of the excited state wave packet for the laser frequency 9000 cm^{-1} (the contours of the density by solid line). The contours of the excited electronic state are superimposed (dotted line). The dashed lines represent the crossing seams induced by the laser field. Reproduced with permission from [3].

other channel due to mode-coupling via the ground electronic state and is finally dissociated into the latter channel. No analytical theory exists for two-dimensional problems, but the one-dimensional theory can be used to some extent in the above-mentioned way to roughly estimate the appropriate conditions. The favorable conditions of selective control in the two-dimensional system may be summarized

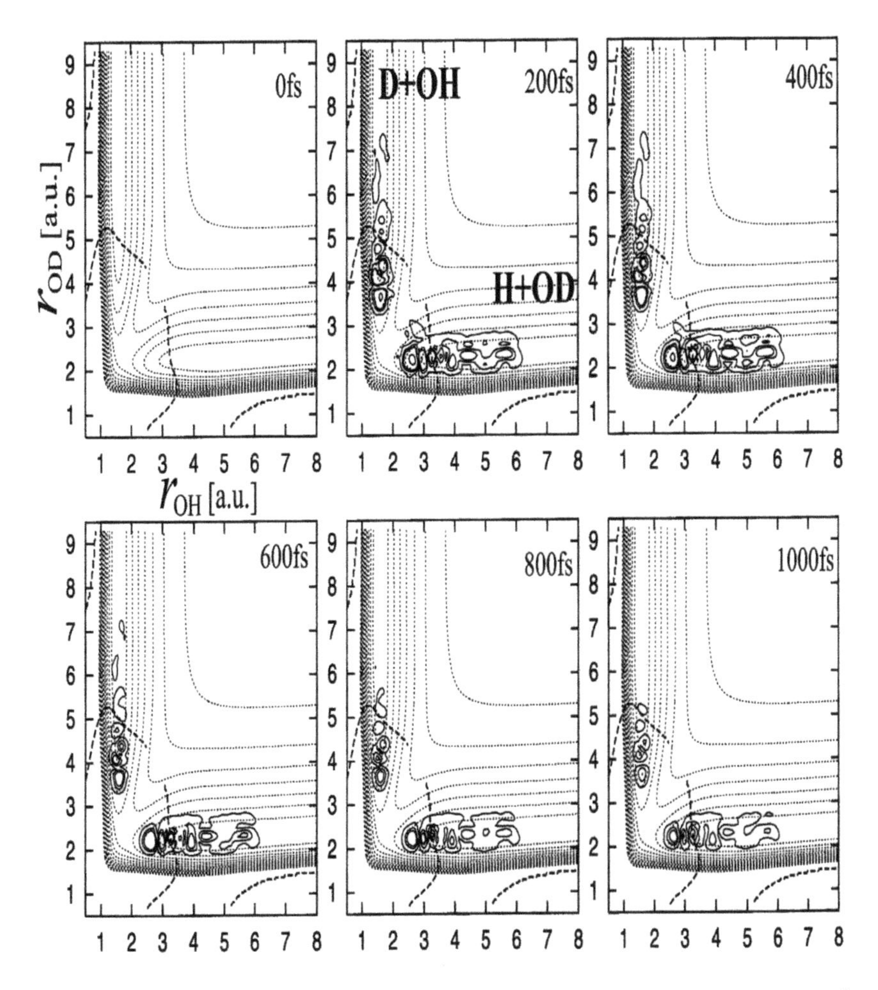

Figure 6.14: The same as Fig. 6.13 for the laser frequency 11500 cm^{-1}. Reproduced with permission from [3].

as follows: The mode-coupling in the ground electronic state should be localized around the equilibrium position and should be negligible in the region of crossing seam created by the laser field. Otherwise the mode-coupling potential destroys the complete reflection condition. In other words, the initial vibrational state should have the one-mode character around the region of crossing seam and the L-shape wave function like that in Fig. 6.10 is favorable.

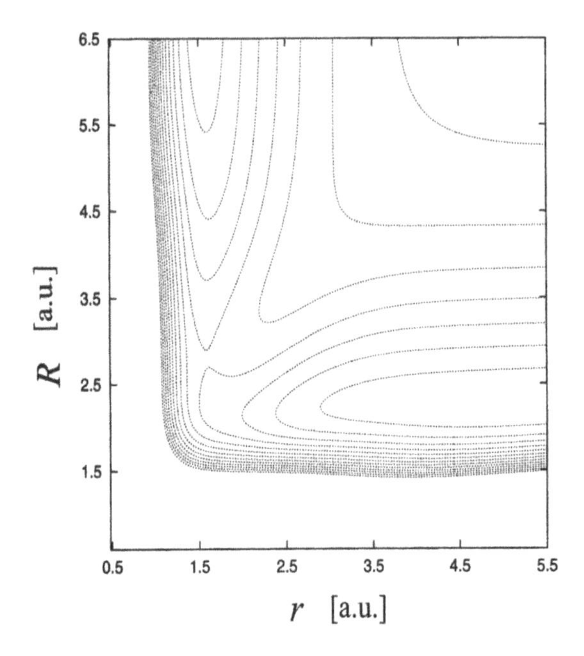

Figure 6.15: Contour plots of the modified excited electronic state of HOD. The saddle point is located in the HO+D channel. The contour spacing is 5000 cm^{-1}. Reproduced with permission from [3].

Our control scheme can also be applied to a two-dimensional system such that the excited electronic state has a potential barrier which is shifted very much to one of the channels and prohibits the dissociation into that channel. In such a model system, we have employed the same potentials used above and slightly modified the excited state potential so that the saddle point is located in the HO+D channel (see Fig. 6.15). The only term of the excited electronic state, $0.2443589 \times 10^2 \times (S_1 S_3^2 + S_1^2 S_3)$ in [10], is changed to $0.1443589 \times 10^2 \times (S_1 S_3^2 + 0.8 S_2^2 S_3)$. With this modification the dissociation into the HO+D channel is not possible anymore along the excited state potential surface, if the initial vibrational state is localized in the H+OD channel. The same initial state and the method of wave packet propagation as those used above are employed. The calculated dissociation probabilities against the laser frequency are shown in Fig. 6.16(a). The solid (dotted) line stands

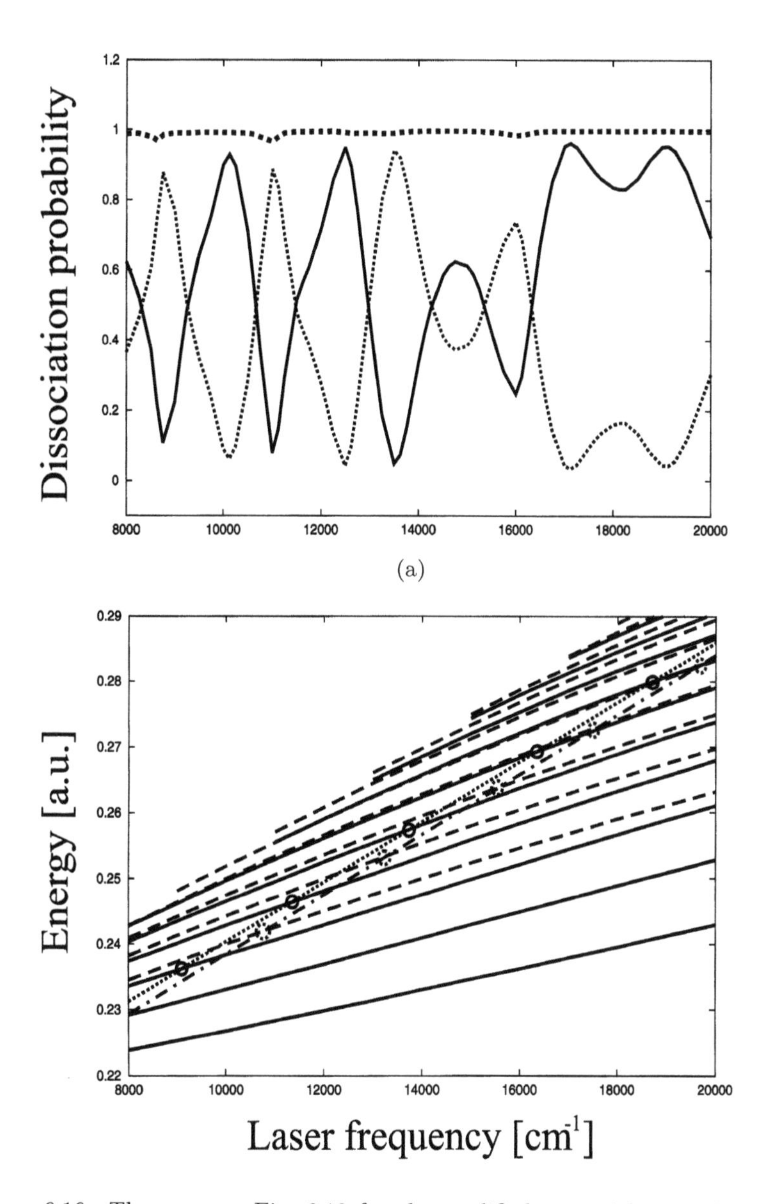

Figure 6.16: The same as Fig. 6.12 for the modified potential as explained in the text.

for the dissociation probability into the H+OD (HO+D) channel. When the laser frequency is either one of $\sim$10000 cm^{-1}, $\sim$12500 cm^{-1}, $\sim$17000 cm^{-1} or $\sim$19000 cm^{-1}, the dissociation into the H+OD channel is preferential, while the dissociation into the HO+D channel is preferential when the laser frequency is set at $\sim$9000 cm^{-1}, $\sim$11000 cm^{-1} or $\sim$13500 cm^{-1}. As is clearly seen, the control is quite selective and the dissociation even into the non-dissociative HO+D channel is possible by adjusting the laser frequency appropriately. The analytical prediction of the complete reflection positions is shown in Fig. 6.16(b) in the same way as in Fig. 6.12(b). The solid (dashed) lines depict the complete reflection positions in the H+OD (HO+D) channel, and the dotted (dash-dotted) line represents the vibrational state $v = 17$ of the O–H bond ($v = 23$ of the O–D bond) shifted up by one photon energy. The solid (dotted) circles show the position of the complete reflection in the H+OD (HO+D) channel. The dissociation probability dips in Fig. 6.16(a) coincide quite well with these analytical predictions. Some of the dips ($\omega \sim$ 16000 cm^{-1} and 18000 cm^{-1}) and the peak at $\omega \sim$ 15000 cm^{-1} are, however, not complete because of the multi-dimensional topography of potential energy surface.

References

[1] H. Nakamura, *Nonadiabatic Transition: Concepts, Basic Theories and Applications*, World Scientific, 2002 (1st edition), 2012 (2nd edition).

[2] H. Nakamura, *Introduction to Nonadiabatic Dynamics*, World Scientific, 2019.

[3] K. Nagaya, Y. Teranishi and H. Nakamura, *J. Chem. Phys.* **113**, 6197 (2000).

[4] J. Fujisaki, Y. Teranishi and H. Nakamura, *J. Theory Comput. Chem.* **1**, 245 (2002).

[5] A.B. Alekseyev, H.P. Liebermann and R.J. Bruenker, *J. Chem. Phys.* **113**, 6174 (2000).

[6] N. Barakrishnan, A.B. Alekseyev and R.J. Bruenker, *Chem. Phys. Lett.* **341**, 594 (2001).

[7] J.E. Stevens, H.W. Jang, L.J. Butler and J.C. Light, *J. Chem. Phys.* **102**, 7059 (1995).

[8] J.G. Izquierdo, G.A. Amarel, F. Ausfelder, F.J. Aoiz and L. Banares, *ChemPhysChem.* **7**, 1682 (2006).
[9] J.R. Reimers and R.O. Watts, *Mol. Phys.* **52**, 357 (1984).
[10] V. Engel, R. Shinke and V. Staemmler, *J. Chem. Phys.* **88**, 129 (1998).
[11] V. Staemmler and A. Palma, *Chem. Phys.* **93**, 63 (1985).
[12] J. C. Light, I. P. Hamilton and J. V. Lill, *J. Chem. Phys.* **82**, 1400 (1985).

Chapter 7

Guided Optimal Control Theory and Directed Momentum Method

7.1 Basic Theory

As explained in Chapter 5, the quadratic chirping method is quite effective for excitation and de-excitation of energy levels and also for pump or dump of wave packet. In the case of vibrational levels of multi-dimensional systems, however, this is not efficient, since the vibrational energy level calculation itself is already very time consuming and actually not feasible. The method is not useful also for controlling the nonadiabatic transition at a naturally existing conical intersection. In this case it is more effective to control the wave packet motion directly so that the nonadiabatic transition at the conical intersection becomes efficient. In this sense, optimal control theory (OCT) is useful. The general idea of OCT is to design such a laser field $\{E_k(t)\}$ that the wave packet $\phi(t)$ propagated from the initial state $\phi(t=0) = \Phi_I$ becomes as close as possible to the desired target state Φ_T at time $t = T$ by solving the appropriate equations iteratively to optimize a certain functional. Various versions of quantum and classical OCTs have been formulated (see the following books and references therein [1, 2]). If we employ the simplest form of global optimization procedure with the iterative gradient search method [3], the correction $\delta E_k(t)$ to the optimal field at each iteration is given by

$$\delta E_k(t) = \hbar^{-1}\langle\Phi_T|\phi(t)\rangle\mathrm{Im}[\Theta_k(t)], \tag{7.1}$$

where the correlation function $\Theta_k(t)$ is defined as

$$\Theta_k(t) = \langle \phi(t) | \mu_k(\mathbf{r}) | \chi(t) \rangle \tag{7.2}$$

and $\mu_k(\mathbf{r})$ is the dipole moment with k denoting the polarization vector component. The wave packets $\phi(t)$ and $\chi(t)$ are propagated forward and backward, respectively, according to the time-dependent Schrödinger equation,

$$\left[\frac{i}{\hbar} \frac{\partial}{\partial t} + \sum_{j=0}^{N} \frac{\hbar^2}{2m_j} \frac{\partial^2}{\partial r_j^2} - V(\mathbf{r}) + \sum_{k=1}^{3} \mu_k(\mathbf{r}) E_k(t) \right] \psi(t) = 0 \tag{7.3}$$

with $\psi(t = 0) = \Phi_I$ in the case of $\psi = \phi$ and $\psi(t = T) = \Phi_T$ in the case of $\psi = \chi$. In the quantum mechanical version of OCT, these Schrödinger equations are solved directly by using the grid method. It is unfortunately very difficult to apply this method to multi-dimensional systems because of the formidable numerical cost. The classical mechanical version, on the other hand, can be easily used for high-dimensional systems. However, they cannot be reliable, since the various phases play crucial roles in the case of laser control of dynamics [4, 5]. Considering these facts, Kondorskiy and Nakamura formulated the semiclassical version of OCT [6–8] with use of the Herman–Kluk-type frozen Gaussian expansion method [9–11]. The wave packet $\phi(\mathbf{r}, t)$ is expanded in terms of the frozen Gaussian functions as

$$\begin{aligned} \phi(\mathbf{r}, t) = & \int_{traj} \frac{d\mathbf{q}_0 d\mathbf{p}_0}{(2\pi)^N} g(\mathbf{r}; \gamma, \mathbf{q}_t, \mathbf{p}_t) C_{\gamma, \mathbf{q}_0, \mathbf{p}_0, t} \exp[\frac{i}{\hbar} S_{\mathbf{q}_0, \mathbf{p}_0, t}] \\ & \times \int d\mathbf{r}_0 g^*(\mathbf{r}_0; \gamma, \mathbf{q}_0, \mathbf{p}_0) \Phi_I(\mathbf{r}_0), \end{aligned} \tag{7.4}$$

where N is the dimensionality of the system and the frozen Gaussians $g(\mathbf{r}; \gamma, \mathbf{q}, \mathbf{p})$ are defined as

$$g(\mathbf{r}; \gamma, \mathbf{q}, \mathbf{p}) = \Pi_{j=1}^{N} \left(\frac{2\gamma_j}{\pi} \right)^{N/4} \exp\left[-\gamma_j (r_j - q_j)^2 + \frac{i}{\hbar} p_j (r_j - q_j) \right], \tag{7.5}$$

where γ_j is a constant parameter common for all wave packets. The action integral S and the pre-exponential factor C are given by

$$S_{\mathbf{q}_0,\mathbf{p}_0,t} = \int_0^t \left[\sum_{j=1}^{N} \frac{p_{j,\tau}^2}{2m_j} - V(\mathbf{q}_\tau) \right] d\tau \tag{7.6}$$

and

$$C_{\gamma,\mathbf{q}_0,\mathbf{p}_0,t} = \sqrt{\frac{1}{2}\mathrm{Det}\left[\frac{\partial \mathbf{q}_t}{\partial \mathbf{q}_0} + \gamma^{-1}\frac{\partial \mathbf{p}_t}{\partial \mathbf{p}_0}\gamma + \frac{i}{2\hbar}\gamma^{-1}\frac{\partial \mathbf{p}_t}{\partial \mathbf{q}_0} - 2i\hbar\frac{\partial \mathbf{q}_t}{\partial \mathbf{p}_0}\gamma \right]}, \tag{7.7}$$

where γ is the diagonal matrix composed of the parameters γ_j. The correlation function Θ is now expressed as

$$\Theta_k(t) = \int_{traj} \frac{d\mathbf{q}_0 d\mathbf{p}_0}{(2\pi\hbar)^N} C^*_{\gamma,\mathbf{q}_0,\mathbf{p}_0,t} \times \exp\left[-\frac{i}{\hbar} S_{\mathbf{q}_0,\mathbf{p}_0,t} \right] \langle \phi(0) | g_{\gamma,\mathbf{q}_0,\mathbf{p}_0} \rangle \Omega_k(\gamma, \mathbf{q}_t, \mathbf{p}_t), \tag{7.8}$$

where

$$\Omega_k(\gamma, \mathbf{q}_t, \mathbf{p}_t) = \int_{traj} \frac{d\mathbf{q}'_0 d\mathbf{p}'_0}{(2\pi\hbar)^N} \langle g_{\gamma,\mathbf{q}_t,\mathbf{p}_t} | \mu_k(\mathbf{r}) | g_{\gamma,\mathbf{q}'_0,\mathbf{p}'_0} \rangle C_{\gamma,\mathbf{q}'_0,\mathbf{p}'_0,t} \times \exp\left[\frac{i}{\hbar} S_{\mathbf{q}'_0,\mathbf{p}'_0,t} \right] \langle g_{\gamma,\mathbf{q}'_0,\mathbf{p}'_0} | \chi(0) \rangle. \tag{7.9}$$

Since the wave functions $\phi(t)$ and $\chi(t)$ are expanded in terms of classical trajectories, the correlation function $\Theta_k(t)$ contains the double summation with respect to these trajectories. This is still very much computationally demanding. We avoid this by taking only those trajectories that run close to each other. This is equivalent to the linearization of classical dynamics:

$$\mathbf{q}'_t \simeq \mathbf{q}_t + \frac{\partial \mathbf{q}_t}{\partial \mathbf{q}_0}\delta\mathbf{q}_0 + \frac{\partial \mathbf{q}_t}{\partial \mathbf{p}_0}\delta\mathbf{p}_0 \tag{7.10}$$

and

$$\mathbf{p}'_t \simeq \mathbf{p}_t + \frac{\partial \mathbf{p}_t}{\partial \mathbf{q}_0}\delta\mathbf{q}_0 + \frac{\partial \mathbf{p}_t}{\partial \mathbf{p}_0}\delta\mathbf{p}_0 \tag{7.11}$$

with $\delta\mathbf{q}_0 = \mathbf{q}_0' - \mathbf{q}_0$ and $\delta\mathbf{p}_0 = \mathbf{p}_0' - \mathbf{p}_0$. The classical action S and the pre-exponential factor C are expanded to the second and the zeroth order, respectively. This approximation is the same as that used in the cellurization procedure or the linearization approximation in semiclassical mechanics [12–16].

In practical calculations it is usually possible to expand the function $\chi(0)$ in terms of the Gaussian functions and the factor $\langle g|\chi\rangle$ can be easily analytically evaluated. Furthermore, if the dipole moment $\mu_k(\mathbf{r})$ can be assumed to be a linear function of $\mathbf{r}$ within the spread of the wave packet, the correlation function can be finally very much simplified as

$$\Theta_k(t) = \int_{traj} \frac{d\mathbf{q}_0 d\mathbf{p}_0}{(2\pi\hbar)^N} \langle\phi(0)|g_{\gamma,\mathbf{q}_0,\mathbf{p}_0}\rangle \times \langle g_{\gamma,\mathbf{q}_0,\mathbf{p}_0}|\chi(0)\rangle [\mu_k(\mathbf{q}_t) - \nabla\mu_k(\mathbf{q}_t)\mathbf{F}_{\gamma,\mathbf{q}_0,\mathbf{p}_0,t}], \quad (7.12)$$

where $\mathbf{F}_{\gamma,\mathbf{q}_0,\mathbf{p}_0,t}$ is a function of trajectory parameters and its detailed expression is given in [7, 8]. If the dipole moment does not change much within the width of the frozen wave packet, then the second term in the square bracket is negligible and the correlation function becomes very simple. This simplifies the numerical calculations of optimal control field very much and the method is applicable to multi-dimensional systems. Another problem of OCT is that the method is very time-consuming and inefficient when the overlap between the initial and target states is small. Some methods to cure this problem have been proposed such as the local control scheme and the use of an intermediate target state [17–19]. The present semiclassical theory cannot be free from this difficulty. To overcome this difficulty, we divide the whole process into a sequence of steps and each step is optimized. In other words, the optimization procedure is performed by setting an appropriate target state in each step. This intermediate target state plays a role to guide the trial state and is not necessarily accurate, since the important factor is the final efficiency. The wave packet obtained as the results of the previous step is used as an initial state for the next step. The whole procedure can be very much accelerated by this method and now enables us to treat high-dimensional systems. Because of the single summation with respect to classical trajectories instead of the double summation, the accuracy of the present method is deteriorated at long-time propagation. The improvement can be made by recalculating the frozen Gaussian expansion coefficients from time to time. The method explained above is called semiclassical *guided* optimal control theory (SCGOCT) [6–8, 20, 21].

Figure 7.1 depicts a simple example of shifting the position of a wave packet [7, 8]. A two-dimensional model of H_2O is used. The upper panel shows the initial Φ_i and the target Φ_t states. These are symmetric Gaussians with full widths at half maximum equal to be 0.5 *a.u.* and zero central momenta and put at $\mathbf{R} = (1.81\ a.u., 1.81\ a.u.)$ and $\mathbf{R} = (2.50\ a.u., 1.81\ a.u.)$, respectively. The optimization procedure runs for 50 fs with a zero initial guess field. The number of trajectories used is 3000. After 8–10 iterations, the changes in the absolute value of the overlap between the controlled wave packet and the target state become compatible with the error of the semiclassical wave packet propagation method itself. Thus, the results obtained after ten iterations are considered to be the final ones. These are shown in the lower panel of Fig. 7.1. The final absolute values of the overlap are 0.96 for the quantum [(b) in the panel] and 0.91 for the semiclassical one with use of the correlation function calculated by Eq. (7.12)[(c) in the panel], and 0.83 with use of the simplified correlation function explained below in Eq. (7.12)[(d) in the panel].

The wave packet momentum vector can also be changed by an appropriate laser field. This is important to control nonadiabatic transitions at naturally existing conical intersections. The periodic chirping method explained in Chapter 5 is effective for the transitions at potential surface intersections among *dressed* states. Namely, these intersections are created artificially by lasers. On the other hand, the conical intersections of molecules playing crucial roles in chemical dynamics are naturally given, and the topography of potential energy surfaces and the nonadiabatic couplings there are determined by Nature and we cannot change them. It is true that the potential energy surfaces can be modified by a strong laser, but such a strong laser induces many other undesirable multi-photon processes and is

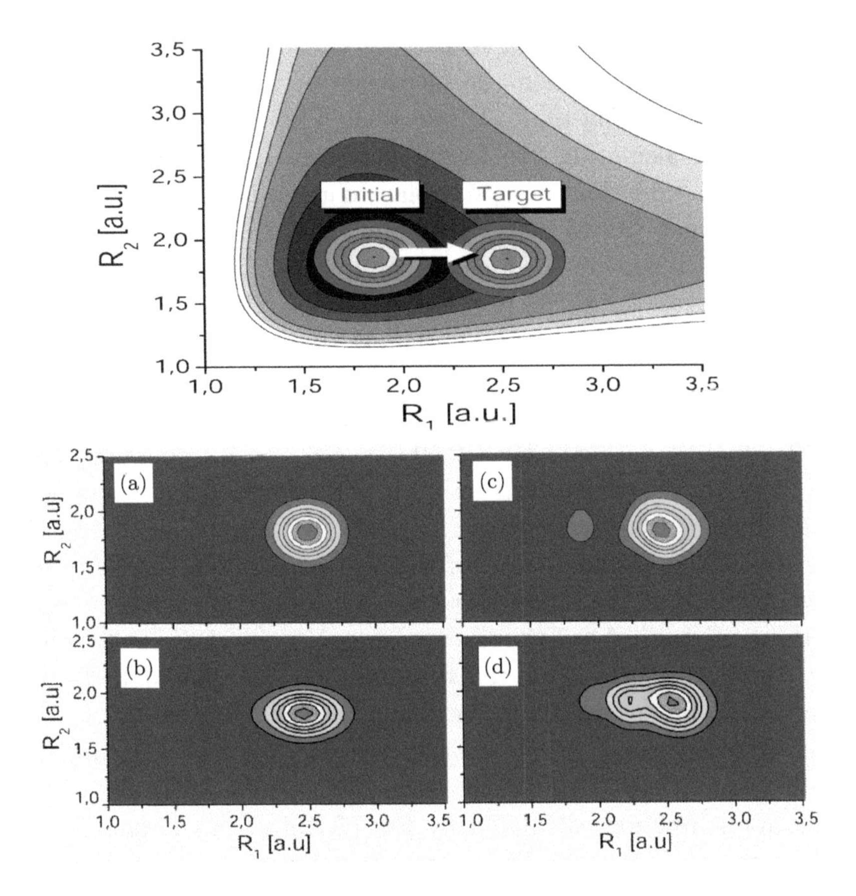

Figure 7.1: Upper panel: ground state potential energy surface of two-dimensional H_2O model system. Circles represent initial Φ_i and target Φ_t states. Lower panel: final wave packets driven by the optimal field calculated quantum mechanically and semiclassically after ten iterations of the optimization algorithm. (a) Target state wave packet, (b) quantum result, (c) semiclassical result with use of the correlation function Eq. (7.12), (d) semiclassical result with use of the simplified correlation function explained below in Eq. (7.12). Reproduced with permission from [7].

not convenient for controlling chemical dynamics in general. Instead, we can design a laser field by using OCT so that the wave packet can gain an appropriate momentum vector to enhance the transition at the conical intersection in a desirable way. This is called the *directed momentum method* [21, 22].

7.2 Photo-Conversion of CHD to HT

Conical intersections in molecules are naturally given, i.e., the topography of the potential energy surfaces and the nonadiabatic coupling between the two surfaces are determined by Nature and we cannot change them. It is true, as mentioned before, that the potential energy surfaces can be modified and the additional coupling between the two surfaces can be induced by a strong laser. However, such a strong laser induces many undesirable multiphoton processes and is not convenient for controlling the transition at the conical intersection. Instead, we can give an appropriate momentum vector to the wave packet so that the nonadiabatic transition at the conical intersection occurs in a favorable way. This is called the *directed momentum method*, as explained in the previous section.

One numerical demonstration of enhancing the photo-conversion efficiency was carried out in the photo-chromic cyclohexadiene (CHD)/hexatriene (HT) system by giving an appropriate directed momentum to the initial wave packet [23]. Giving an appropriate momentum to the wave packet is just an example of controlling wave packet motion and the optimal laser can be designed by using SCGOCT. A very rough consideration about the appropriate laser for directing the momentum vector may be made as follows. The momentum given to the wave packet by laser is expressed as follows [25]:

$$P = -i\hbar\langle\Psi_0|\hat{U}^\dagger(t)\nabla_{\mathbf{r}}\hat{U}(t)|\Psi_0\rangle - i\hbar\langle\Psi_0|\hat{U}^\dagger(t)(\nabla_{\mathbf{r}}\hat{\Omega}(t))\hat{U}(t)\Psi_0\rangle \tag{7.13}$$

with

$$\hat{\Omega}(t) = \int_0^t \hat{H}(\tau)d\tau + \frac{1}{2}\int_0^t d\tau \int_0^\tau d\tau' \big[\hat{H}_I(\tau) + \hat{H}_0(\tau), \hat{H}_I(\tau') + \hat{H}_0(\tau')\big] + \cdots. \tag{7.14}$$

Here $\mathbf{r}$, $|\Psi_0\rangle, \hat{U}(t) = \exp[\hat{\Omega}(t)]$, $\hat{H}_0$, and $[\cdot\cdot, \cdot\cdot]$ are the molecular internal coordinates, the initial state, the evolution operator of the system, Hamiltonian of the molecule, and the commutator, respectively.

The laser-molecule interaction is $H_I = -\vec{\mu}(\mathbf{r}) \cdot \mathbf{E}(t)$, where $\vec{\mu}(\mathbf{r})$ and $\mathbf{E}(t)$ are the 3-D dipole moment vector and laser field vector, respectively. Since $\nabla_{\mathbf{r}}\hat{\Omega}(t)$ contains $\nabla_{\mathbf{r}}\vec{\mu}(\mathbf{r}) \cdot \mathbf{E}(t)$, it is expected that the appropriate laser field is parallel to the derivative of the dipole moment vector in the direction of the desirable momentum. In the case of a polyatomic molecule, the number of molecular internal coordinates is larger than the degrees of freedom of the laser field, but the important geometry of conical intersection is usually determined locally by a few molecular internal coordinates and it is expected to be feasible to adjust the laser field to the appropriate direction, as mentioned above.

Photo-induced reversible ring opening/closure of photo-chromic molecules such as diarylethenes and fulgides are applicable to molecular switches and memories [26–29]. Murakami *et al.* [28] have found that the reaction yield of the photo-induced ring opening of diarylethene strongly depends on the time duration of laser irradiation. Control of the reaction yield of photo-chromism by the optimal laser is thought to be particularly important for the "gated function", which is crucial to achieve low fatigue and non-destructive readout capability in practical photo-switches and optical rewritable memories (i.e., efficiently isomerized by the specific optimal laser, but not isomerized by sunlight or readout light) [28]. The clarification of the ultrafast photochemical dynamics enables us to build up reaction control strategies to achieve high isomerization yield, quick response, and gated functions, which are required in practical photo-chromic systems. Recent advances in experimental and theoretical techniques have made it possible to reveal time-dependent pictures of ultrafast reactions in detail [30], which provides a wide perspective for reaction controls and molecular designs. Photo-isomerization between CHD and HT has been attracting much interest [31–36] not only as a prototype of ultrafast photochemistry but also as a model system to understand universal reaction mechanisms of photo-chromism, because CHD/HT is the reaction center of various photo-chromic molecules. The ground state CHD equilibrium geometry is of C_2 symmetry, at which the electronic ground (S_0), first excited (S_1) and second excited (S_2) states have $1^1A, 1^1B$, and 2^1A characters, respectively, and the photo-excitation to $S_1(1^1B)$ is much stronger than that to $S_2(2^1A)$ because of large $1^1A - 1^1B$ transition dipole moment. Fuß *et al.* [31] have experimentally observed the CHD/HT photo-isomerization in vapor phase in femtosecond time resolution as follows (with time durations from photo-excitation):

(i) Photo-excitation of CHD to $S_1(1^1B)$ in the Franck-Condon region (0 fs).
(ii) State character change of S_1 from 1^1B to 2^1A (53 fs).
(iii) Radiationless decay from S_1 to S_0 (130 fs).
(iv) CHD or HT formation on S_0 (200 fs).

Tamura *et al.* studied this CHD/HT photo-isomerization process as a prototype system of photo-chromism [21, 23, 24, 34]. They calculated the relevant potential energy surfaces along the two-dimensional coordinates shown in Fig. 7.2(a). The other non-reactive coordinates are optimized. The single-point energies of the relevant states at the optimized geometries are calculated by using the multi-reference configuration interaction method with single excitations to take into account dynamic correlation energies. The geometrical structure of the minimum of the conical intersection between $S_1(2^1A)$ and $S_0(1^1A)$ is shown in Fig. 7.2(b). The reaction scheme is shown in Fig. 7.3. First, the photo-excitation occurs from the ground state $S_0(1^1A)$ to $S_1(1^1B)$, since the transition dipole moment to $S_1(1^1B)$ is much larger than that to $S_2(2^1A)$. There are two important conical intersections on the way: one is between S_1 and S_2 and the other

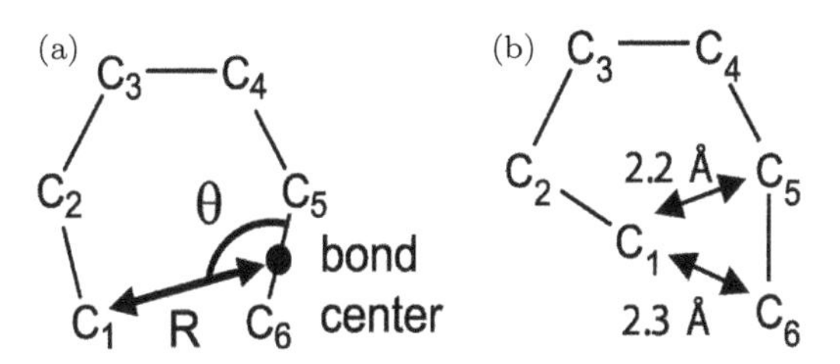

Figure 7.2: (a) Two-dimensional Jacobi coordinates used in the wave packet dynamics. (b) Schematic molecular geometry at the S_1–S_0 conical intersection minimum (minimum on the conical intersection hypersurface). Reproduced with permission from [24].

between $S_1(2^1A)$ and $S_0(1^1A)$. At the first conical intersection, the state character changes as $S_1(1^1B) \rightarrow S_1(2^1A)$ and $S_2(2^1A) \rightarrow S_2(1^1B)$. This occurs by breaking the C_2 symmetry. The two-dimensional wave packet dynamics calculations are performed in the diabatic representation with use of these potential energy surfaces. The reduced mass is taken as that of the CH_2–ethylene system. The dynamics of the natural (not controlled) CHD/HT photo-isomerization can be summarized as follows (see Figs. 7.3 and 7.4) [21, 23, 24]:

(i) The initial wave packet (vibrational ground state on the electronic ground S_0 state in the CHD region) is photoexcited to the $S_1(1^1B)$ electronic state and then descends that potential energy surface toward the open-ring direction (steps (1) and (2) in Fig. 7.3).

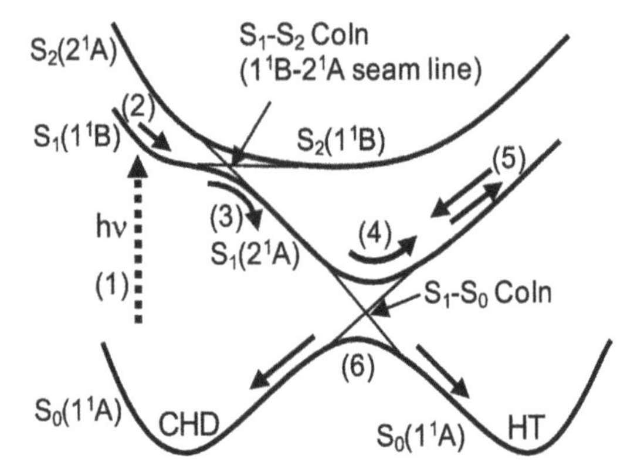

Figure 7.3: Reaction scheme of CHD/HT photoisomerization. (1) CHD is photoexcited to $S_1(1^1B)$ (0 fs), (2) ring opening (C–C bond breaking) proceeds descending the $S_1(1^1B)$ surface, (3) electronic state character of S_1 changes from 1^1B to 2^1A at the 1^1B–2^1A seam line (~20 fs), (4) $S_1(2^1A)$ wave packet ascends the potential energy surface toward the open-ring direction due to the excess kinetic energy from the Franck-Condon region, where the wave packet is still compact and does not spread over the S_1–S_0 conical intersection, (5) wave packet turns back toward the closed-ring direction (~70 fs), and (6) is scattered by the steep potential slope at the closed-ring (100–130 fs), and then radiationless decay to S_0 occurs (130–180 fs) through the S_1–S_0 conical intersection along the direction toward the 5MR. Reproduced with permission from [21].

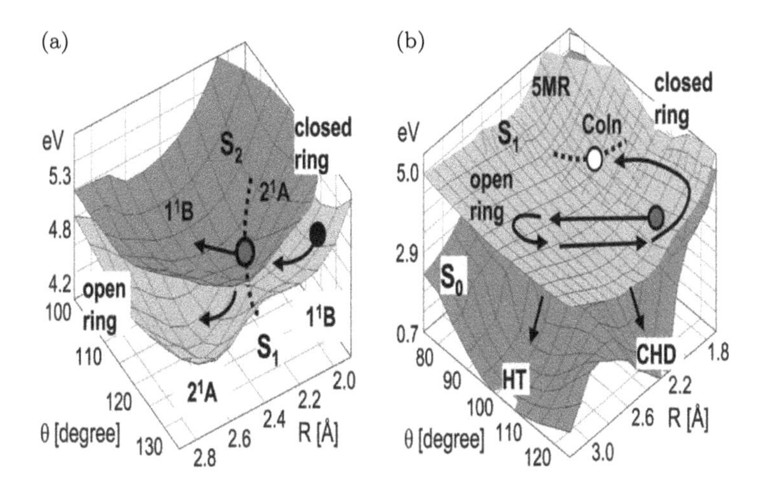

Figure 7.4: (a) S_1–S_2 and (b) S_1–S_0 coupled potential energy surfaces along the two-dimensional Jacobi coordinates. The black, gray, and white circles and dotted lines indicate the locations of the Franck-Condon region, the S_1–S_2 conical intersection minimum, 5MR S_1–S_0 conical intersection minimum, and seam lines, respectively. The solid arrows indicate the schematic wave packet pathway of natural photo-isomerization starting from the vibrational ground state. Reproduced with permission from [24].

(ii) This wave packet on the $S_1(1^1B)$ state encounters the S_1–S_2 conical intersection located along the C_2 symmetric pathway, at which the major portion (~80%) changes the character from 1^1B to 2^1A, staying on the S_1 adiabatic state (see Fig. 7.4(a)).

(iii) The $S_1(2^1A)$ wave packet goes to the open-ring region from the Franck-Condon region and turns back toward the closed-ring region due to the gently ascending potential slope (Fig. 7.4(b)), where the wave packet is still compact and does not spread over the S_1–S_0 conical intersection.

(iv) The wave packet is scattered by the steep potential slope at the closed-ring region and reaches the S_1–S_0 conical intersection located along the direction toward the five-membered ring (5MR) (see Fig. 7.4(b)).

(v) Finally, the first nonadiabatic transition to S_0 occurs at the S_1–S_0 conical intersection and the S_0 wave packet bifurcates into the CHD and HT regions (step (6) in Fig. 7.3).

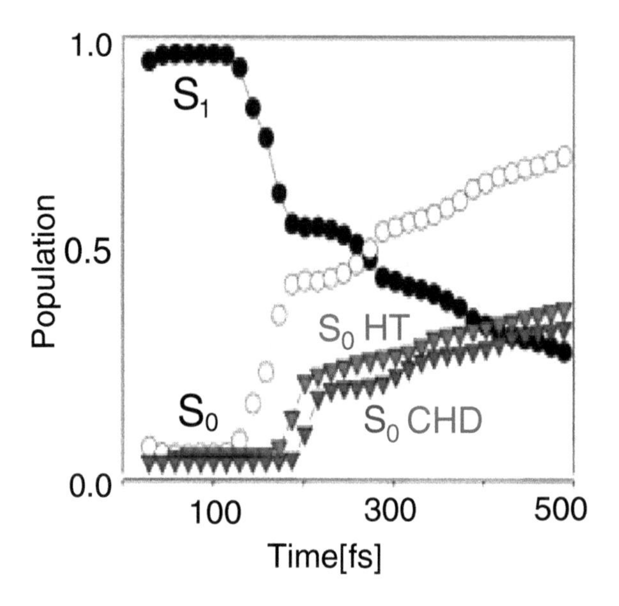

Figure 7.5: Time variation of wave packet populations on S_0 and S_1 states after the initial excitation. Reproduced with permission from [24].

This reaction dynamics reasonably explains the experimentally observed ultrafast $S_1 \rightarrow S_0$ decay (130 fs) [31]. The CHD:HT branching ratio after the $S_1 \rightarrow S_0$ decay is approximately ~5:5 in the wave packet dynamics [21, 23, 24, 37] (see Fig. 7.5). This is because the S_1–S_0 conical intersection (Fig. 7.2(b)) is located at the halfway ridge between the CHD and HT basins on the S_0 potential energy surface (see Fig. 7.4(b)), where the wave packet is rather delocalized when crossing the S_1–S_0 conical intersection compared to the initial one in the Franck-Condon region. The experimental CHD:HT branching ratio is ~6:4 in solution [38]. The results of the wave packet dynamics are generally consistent with the experiments [31, 38].

On the basis of the wave packet dynamics mentioned above, the following control scheme can be proposed to increase the photoisomerization efficiency by (i) preparing the initial wave packet with the directed momentum toward the 5MR in the CHD potential basin on the S_0 electronic state and (ii) achieving nearly complete electronic excitation to $S_1(1^1B)$ (step (1) in Fig. 7.3) with use of the quadratically chirped pulse (see Fig. 7.6) [21, 24]. The directed momentum method enables us to control the motion of the wave packet at the S_1–S_0 conical intersection and enhance the HT/CHD branching ratio. In order to calculate the controlling laser field to prepare the initial wave packet with an appropriate momentum vector in the CHD potential basin on the S_0 electronic state, the method of SCGOCT explained above is used. The initial state is the ground vibrational state at the potential minimum of S_0 state in the CHD configuration. The target state is set to be a Gaussian wave packet with the same central coordinates and width parameters as those of the initial wave packet, but with different central momenta $\mathbf{P}_R = -16.21$ and $\mathbf{P}_\theta = -39.25$ in atomic units. This corresponds to the ~6 kcal/mol kinetic energy directed toward the 5MR. The acceleration duration is set to be 400 fs. The efficiency of control achieved is ~93%. This accelerated wave packet can be excited to the $S_1(1^1B)$ state by using the quadratically chirped pulse with efficiency as high as ~90%. The laser intensity required is about 3.5 TW/cm^2. The laser frequency is chirped as

$$\omega(t) = \alpha_\omega(t - t_c)^2 + \beta_\omega, \tag{7.15}$$

where $\alpha_\omega = 0.07$ eV/fs^2, $\beta_\omega = 4.6$ eV, and $t_c = 6.0$ fs are used. Thus excited wave packet on $S_1(1^1B)$ state goes toward the 5MR due to the initial directed momentum, even though the $S_1(1^1B)$ potential energy surface is downhill along the C_2 symmetric ring-opening direction (see Fig. 7.4(b) without control). The electronic state character changes to $S_1(2^1A)$ at the 1^1B–2^1A seam line (step (3) in Fig. 7.3). Since the wave packet pathway toward the 5MR is far from the S_1–S_2 conical intersection located on the C_2 symmetric pathway (Fig. 7.4(a)), the nonadiabatic transition to $S_2(1^1B)$ is negligible. The S_1 wave packet directly goes to the S_1–S_0 conical intersection due to the initial momentum vector directed toward the 5MR without any excursions around the open-ring region (i.e., skipping the steps (4) and (5) in Fig. 7.3) (see also Fig. 7.6). Figure 7.7 shows the time variation of the wave packet populations on S_1 and S_0 states and the branching ratio. The first nonadiabatic transition to S_0

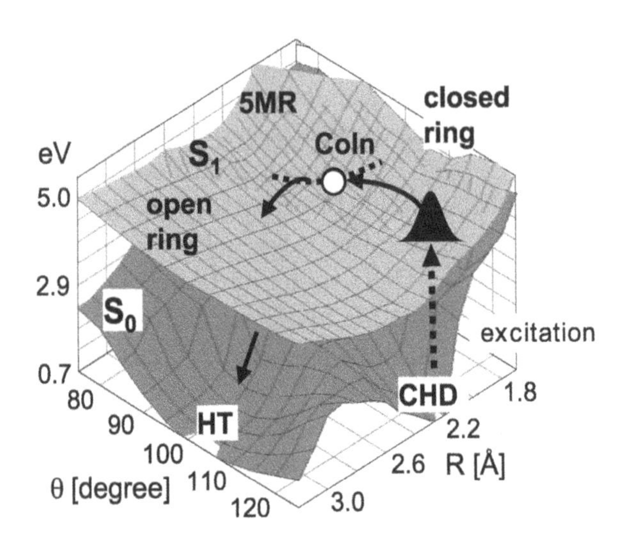

Figure 7.6: The *controlled* wave packet pathway on the S_1–S_0 coupled potential energy surfaces. The white circle and dotted line indicate the locations of 5MR S_1–S_0 conical intersection minimum and seam line, respectively. The solid arrows indicate the wave packet motion toward the 5MR. Reproduced with permission from [21].

rapidly occurs in 20–30 fs after the photo-excitation, keeping the wave packet compact. The $S_1$and S_0 potential energy surfaces are uphill toward the 5MR and thus the wave packet turns around toward the CHD or HT product region. After the $S_1 \rightarrow S_0$ decay at the conical intersection, most of the S_0 wave packet goes down to the HT product side, because the group velocity of the wave packet is directed toward HT. Although the 5MR S_1–S_0 conical intersection is located at the halfway ridge between the CHD and HT potential basins, the desired HT product can be selectively produced because of the desirable momentum vector at the conical intersection. In comparison with Fig. 7.5, Fig. 7.7 clearly demonstrates the increase of HT production by the present scheme. The optimal laser field is confirmed to be parallel to the derivative of the dipole moment vector in the direction of the desirable momentum. The present computations clearly demonstrate the general theoretical possibility of enhancing the photo-conversion efficiency by using the quadratic chirping method and the directed momentum method.

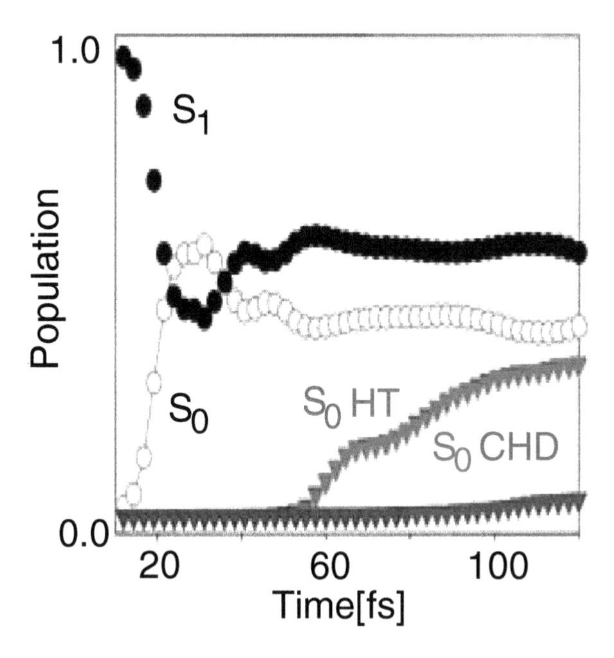

Figure 7.7: Time variation of wave packet populations on S_0 and S_1 states after the initial excitation with use of the *directed momentum method*. Reproduced with permission from [37].

7.3 Selective Photo-Dissociation of OHCl

The bond-selective photodissociation of OHCl to O + HCl can be achieved with high efficiency by the following procedures [21]: (i) appropriately accelerate the ground state wave packet using the method of SCGOCT and (ii) excite thus-prepared wave packet from the ground ($1A'$) to the excited ($2A'$) state using a quadratically chirped laser pulse.

We invoke the non-rotating approximation based on the fact that the time scale of control of about 1.0 ps is much less than the rotation period of the molecule, which is about 32 ps. This provides us with the following 4D model in which the rotation of the whole molecule is allowed in the molecular plane. The molecule is described in terms of two two-dimensional vectors: $\mathbf{R}_{O=Cl}$ for the vector from O to Cl and $\mathbf{R}_H$ for the vector from the center of mass of OCl to H. The space-fixed Cartesian framework is used with X-axis set to be parallel to the initial direction of $\mathbf{R}_{O=Cl}$, and the Y-axis is perpendicular to it. The centers of mass of OCl and of the whole system are assumed to be the same so that the kinetic energy part of the Hamiltonian is diagonal. The potential energy surface and dipole moment are taken from [39]. The ground state wave packet is approximated by a Gaussian function centered (in atomic units) at $\mathbf{R}_{O=Cl} = (3.2633,\ 0.0)$ and $\mathbf{R}_H = (-2.5726,\ 1.7954)$. The width parameters are estimated by fitting the principal modes of the potential wells. The potential energy surfaces of $1A'$ and $2A'$ states are presented in Fig. 7.8.

In order to achieve the process (i) mentioned above, we set the target state to be a Gaussian wave packet with the same central coordinates and width parameters as those of the initial wave packet, but with the central momenta of hydrogen $\mathbf{P}_H = (16.8,\ 2.2)$ in atomic units and zero momenta of the OCl bond, as illustrated in Fig. 7.8. The target state is designed so that after the almost complete electronic excitation to $2A'$ state the wave packet would start to move towards the OClH configuration. The acceleration is necessary, since on the electronic $2A'$ state the HOCl and OClH configurations are separated by a potential barrier.

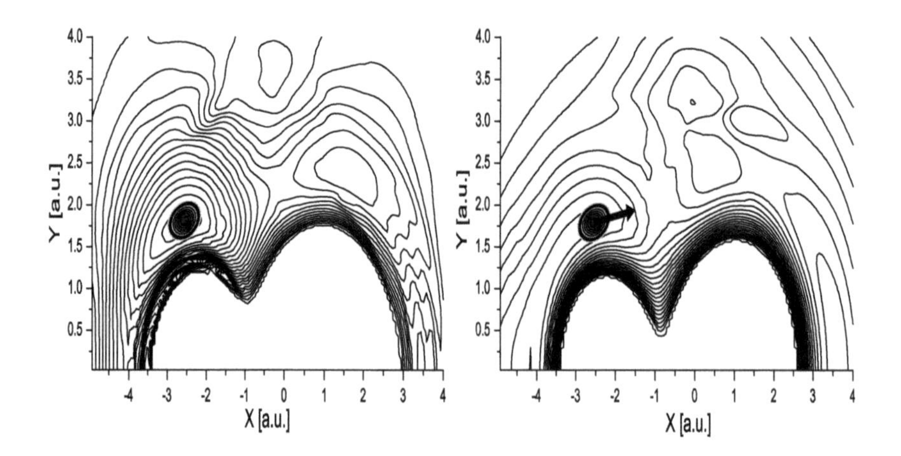

Figure 7.8: Potential energy surfaces of $1A'$ (left) to $2A'$ (right) states of OHCl as functions of proton coordinate together with ground state wave packet. Left: $1A'$ potential energy surface and ground state wave packet. Right: The arrow shows the direction of acceleration. The X-axis is parallel to the initial direction of the O–Cl bond. Reproduced with permission from [21].

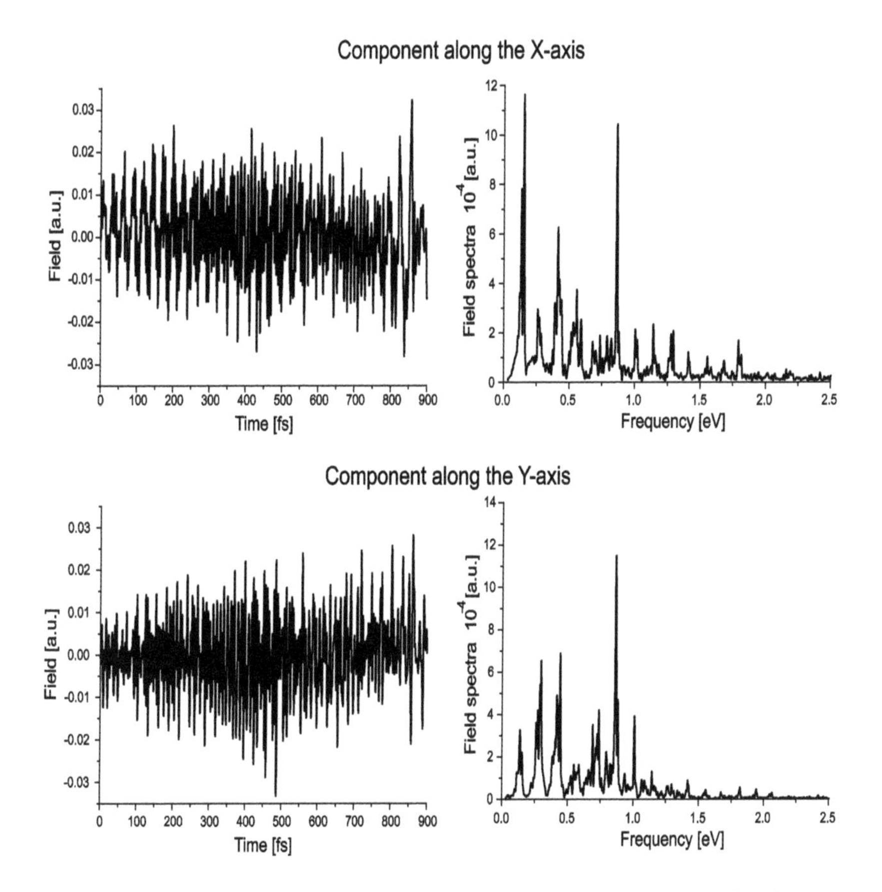

Figure 7.9: Optimal field found for OHCl wave packet acceleration (left) and its spectrum (right). Reproduced with permission from [21].

The controlling time for OHCl wave packet acceleration is set to be 900 fs. The final overlap between the controlled wave packet and the target state achieved after 10 iterations is 85%. Figure 7.9 presents the controlling laser field.

After the wave packet gains the directed momentum, the almost complete electronic excitation to the $2A'$ state by quadratically chirped laser pulse is performed. The parameters of the pulse are: $\alpha_\omega = 3.3 \cdot 10^{-5}$ eV/fs^2, $\beta_\omega = 5.17$ eV, and the intensity $=$ 4.25 TW/cm^2. The resultant wave packet starts to propagate on the excited potential energy surface as shown in Fig. 7.10 where the

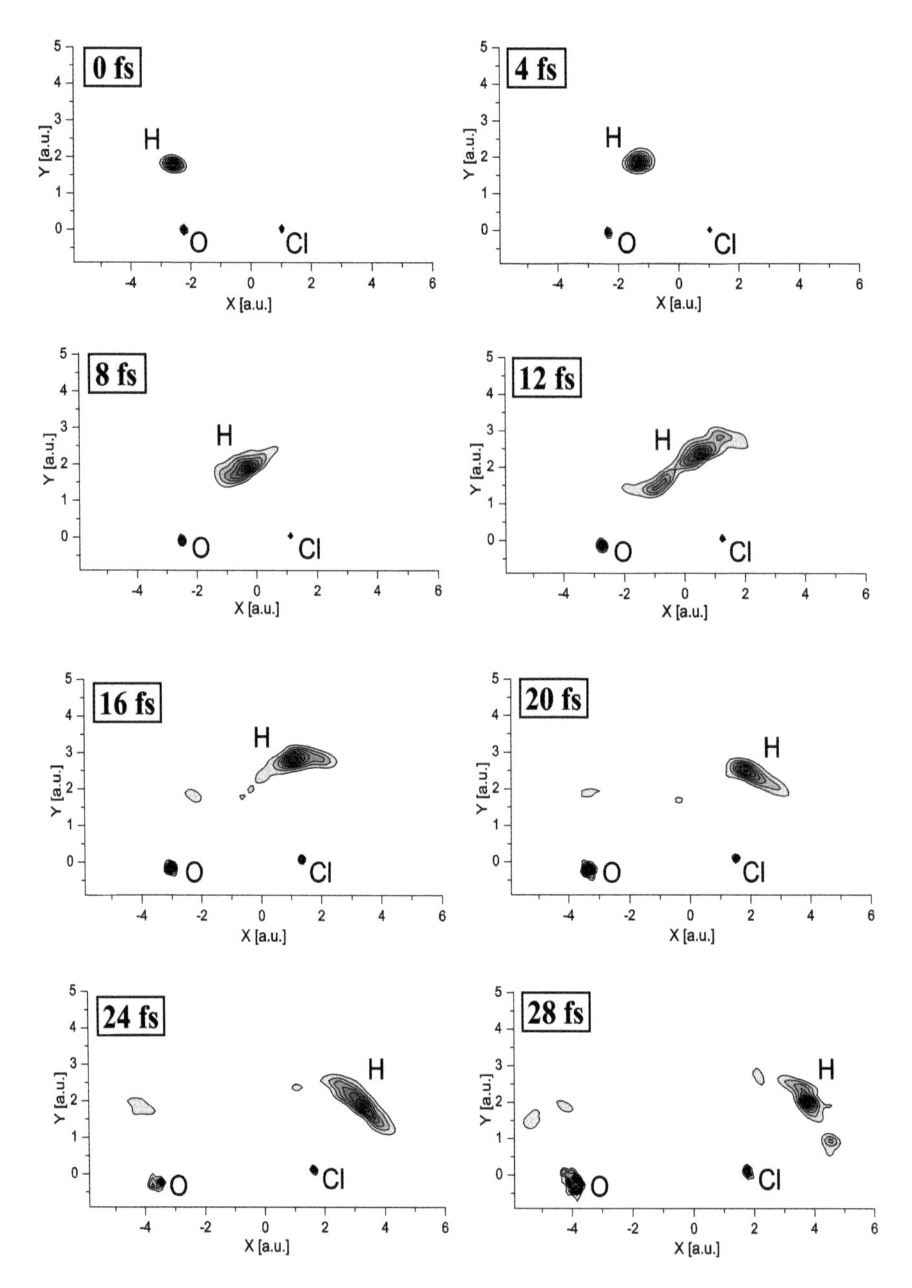

Figure 7.10: Snapshots of the wave packets of H, O and Cl at various moments after the electronic excitation of the accelerated ground state wave packet. The X-axis is parallel to the initial direction of the O–Cl bond. Reproduced with permission from [21].

snapshots are depicted until the final dissociation into the O+HCl channel. The final dissociation probability is about 92% compared to the almost zero probability in the ordinary photo-dissociation.

As mentioned at the end of Section 7.2, the controlling laser field is parallel to the derivative of the dipole moment vector in the direction of the desirable momentum.

7.4 Vibrational Isomerization of HCN

It is demonstrated here that the present SCGOCT makes it possible to realize the full-dimensional or six-dimensional treatment of the isomerization of HCN→CNH [6, 21, 40]. The isomerization or dissociation dynamics of HCN has been considered either with only a small number of degrees of freedom taken into account [41, 42] or by using purely classical methods (see [43] and references therein). The process of NCH-HNC isomerization occurs without change of electronic state and thus what we have to control is the nuclear motion on a single adiabatic potential energy surface together with the molecular rotation. In order for the calculations to be realistic the intensity of the controlling field should not be so high as to cause tunneling ionization, multiphoton electronic excitation, or dissociation. The maximal value of the intensity of the laser pulse which could be used to manipulate the HCN molecule without any damage through ionization in the IR region is estimated to be $\sim 10^{14}$ W/cm^2 [41], which corresponds to the field amplitude of 0.05(atomic units). This requirement results in a longer controlling duration and the rotation of molecule as a whole should be controlled at the same time. Thus the present treatment is a full six-dimensional calculation.

The HCN molecule is described in terms of two three-dimensional vectors [40]: $\mathbf{R}_{N=C}$ for the vector from N to C and $\mathbf{R}_H$ for the vector from the center of mass of CN to H. The space-fixed Cartesian framework is used with X-axis set to be parallel to the initial direction of $\mathbf{R}_{N=C}$, and Y- and Z-axes are perpendicular to it. The centers of mass of NC and of the whole system are assumed to be the same so that the kinetic energy part of the Hamiltonian is diagonal. The molecule is assumed to be aligned after some laser-induced process [41]. The potential energy surface and dipole moment are taken from [44]. The ground state wave packets for the HCN and CNH configurations are approximated by Gaussian functions centered (in atomic units) at $\mathbf{R}_{N=C} = (2.1785, 0.0, 0.0)$ and $\mathbf{R}_H = (3.1855, 0.0, 0.0)$, and $\mathbf{R}_{N=C} = (2.197, 0.0, 0.0)$ and $\mathbf{R}_H = (-2.875, 0.0, 0.0)$, respectively. The width parameters are calculated to fit the principal modes of the potential wells.

According to the idea of guided control, the whole process is divided into the following two main procedures [6, 40]: (i) acceleration of the initial wave packet in the HCN configuration so that it can pass the interstate barrier, and (ii) deceleration of the wave packet so that it stays in the target region. The field intensity is monitored at each step and if its maximum amplitude exceeds the value of 0.05(atomic units), it is reduced to that value by multiplying an appropriate factor. The spectrum of controlling field typically contains one main peak and several harmonics with lower intensity. These secondary peaks are adjusted by this intensity reduction procedure. When the intensity is thus reduced, the controlling time duration should also be changed to maximize the control efficiency. By repeating computations for different time durations T, we can find the best one. The shortest duration for the maximum efficiency under the intensity lower than the ionization threshold is finally found to be 288 fs, where the acceleration lasts for 168 fs and the deceleration lasts for 120 fs.

The acceleration procedure in turn is further divided into the following three steps with intermediate target states of Gaussian wave packets with the same central coordinates and width parameters as those of the initial wave packet, but with the following different central momenta for each step (in atomic units): $\mathbf{P}_{N=C} = (0.0, -15.0, 0.0)$ and $\mathbf{P}_H = (0.0, 10.0, 0.0)$ for the first step, $\mathbf{P}_{N=C} = (0.0, -22.5, 0.0)$ and $\mathbf{P}_H = (0.0, 15.0, 0.0)$ for the second, and $\mathbf{P}_{N=C} = (0.0, -29.2, 0.0)$ and $\mathbf{P}_H = (0.0, 20.0, 0.0)$ for the third. The optimization procedure for the first step is carried out with zero initial guess field. The field obtained as a result of the previous step is used as an initial guess field for the next.

Similarly, in order to decelerate the wave packet so that it stays in the CNH target configuration potential well, the three-step optimization procedure is used, but in a slightly different way. The initial guess field is zero for each step and the wave packet obtained as a result of the previous step is set to be an initial state for the next. The actual wave packet which can energetically pass over the CNH potential well should be decelerated by the first step so that it hits the opposite side of the potential well and starts to move backward. We set the first intermediate target state to be the same as the ground state wave packet in the potential well corresponding to CNH with counterclockwise rotation for a certain angle. This angle and time duration of the first step is adjusted so that the wave packet is decelerated enough to stay in CNH potential and the controlling field intensity is lower than the ionization threshold. The suitable values found are 26° for angle and 40 fs for the corresponding duration. The intermediate target states for the second and third steps are set to be the ground state wave packet in the potential well corresponding to CNH turned clockwise for 26°. These states are set to follow clockwise rotation of the heavy NC bond which occurs during the isomerization. Each step takes 40 fs.

The number of trajectories used at each step of control is kept at 100 000. Typically, the convergence is achieved by about five iterations at each step. The final isomerization probability is calculated by the formula

$$P_{iso} = \langle \phi(T)| \, h\,(0.35 - \cos\theta) \, |\phi(T)\rangle, \tag{7.16}$$

where θ is the angle between $\mathbf{R}_{N=C}$ and $\mathbf{R}_H$, $h(x) = 1$ for $x \geq 0$ and 0 for $x < 0$, and $\cos\theta = 0.35$ corresponds to the saddle point between the HCN and CNH configurations. The final isomerization efficiency attained is 74%. It should be noted, however, that the time average of the field turns out to be non-zero, which means that the zero-frequency component exists. The field with this zero-frequency component removed and its spectrum are shown in Fig. 7.12. Since the system remains symmetric along the XY plane, it has no dipole moment in the direction of the Z-axis. As a result, the Z-component of the field is zero. The snapshots of the wave packet driven by

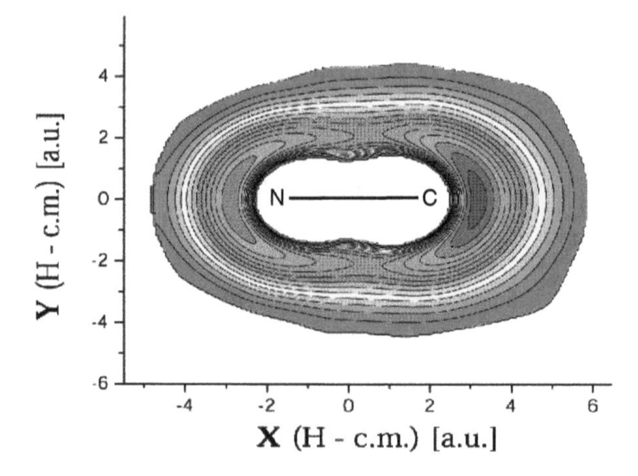

Figure 7.11: Potential contour of HCN with C–N bond fixed at equilibrium position.

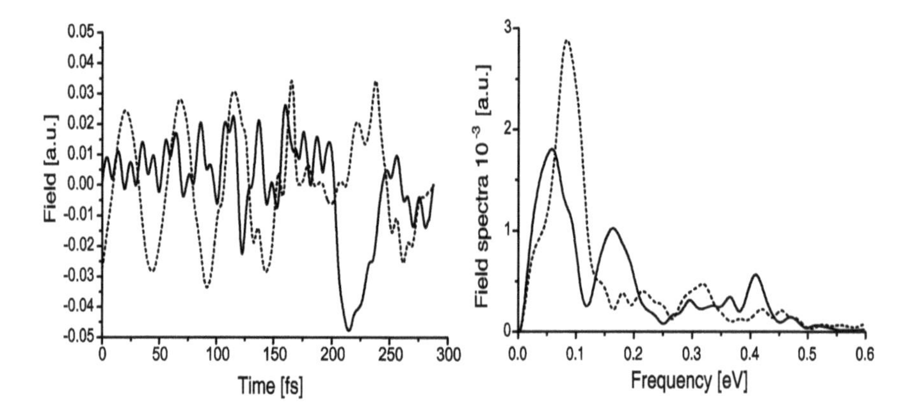

Figure 7.12: Optimal field with the zero frequency component removed (left) and its spectra (right). Solid line: component along the X-axis; dash line: component along the Y-axis. Reproduced with permission from [21, 40].

the controlling field at various times are shown in Fig. 7.13 for the probability density of both proton and N=C bond. The four crucial steps of proton motion are shown in Fig. 7.14: (a) The first step of control provides an appropriate momentum to the proton and moves it, (b) the second step moves the proton over the potential barrier, (c) the third is to freely propagate the proton towards the nitrogen

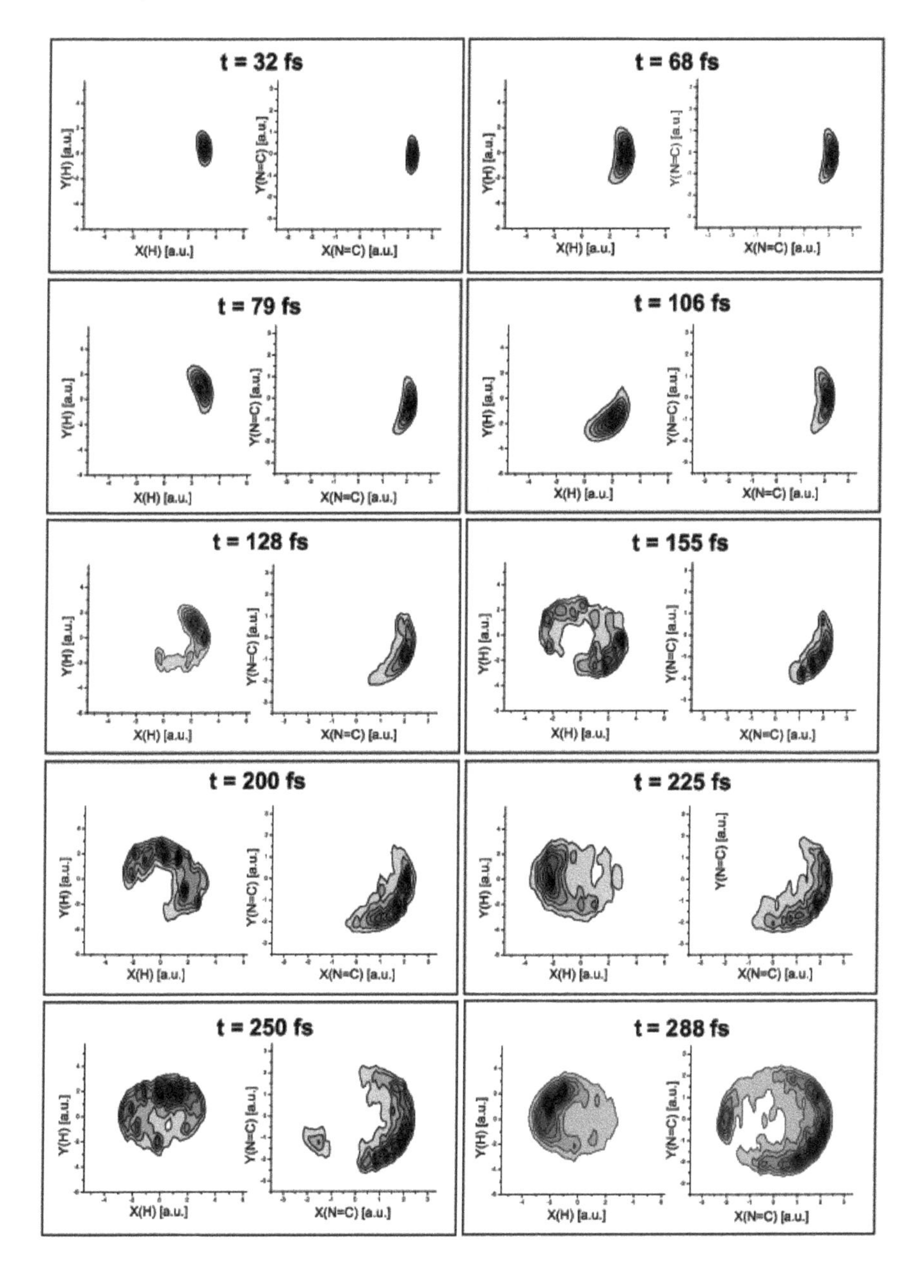

Figure 7.13: Snapshots of the wave packet in the HCNH-CNH isomerization driven by the controlling field at various times. The probability density is given at each figure as a function of proton coordinate (left) and N–C bond vector coordinate (right). Reproduced with permission from [21].

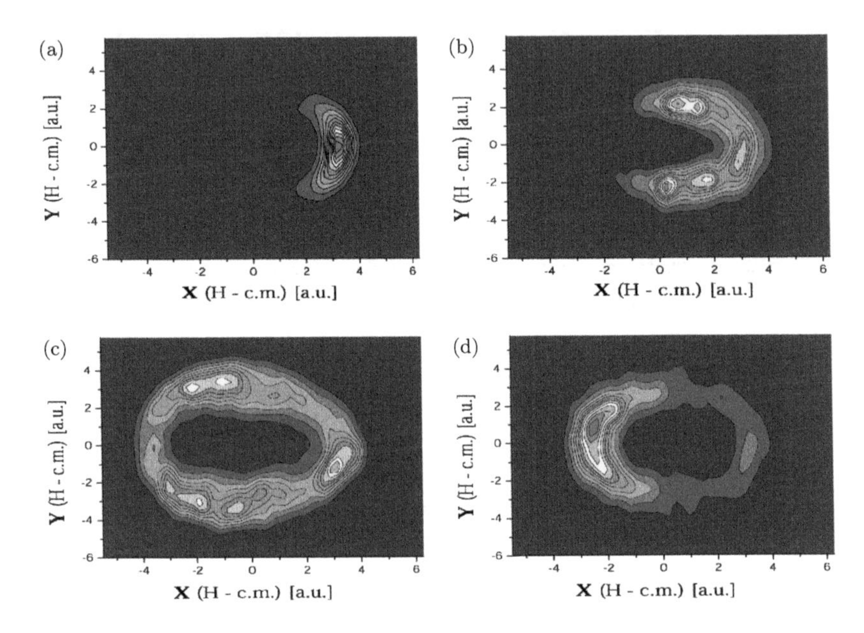

Figure 7.14: Two-dimensional projection of snapshots of the wave packet in the CNH-HCN isomerization driven by the controlling laser field. The probability density is given at each figure as a function of proton coordinate: (a) first step of control to provide an appropriate momentum to the proton, (b) second step of control to overcome the barrier, (c) free propagation towards the nitrogen side, and (d) 30 fs after the free propagation. Reproduced with permission from [6, 40].

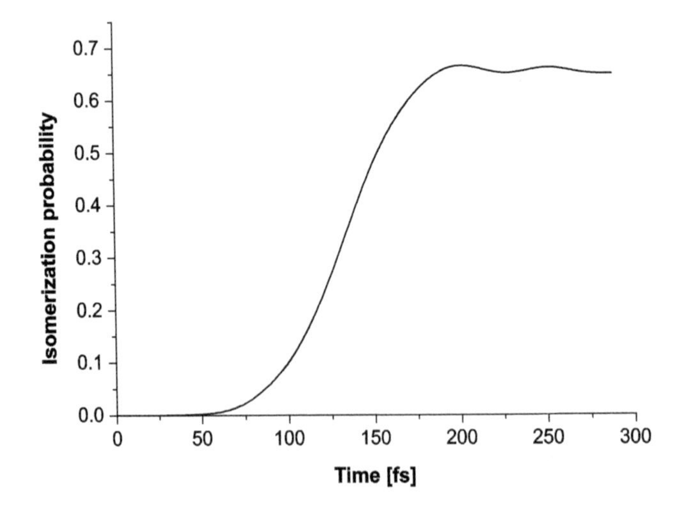

Figure 7.15: HCN-CNH isomerization probability as a function of time. Reproduced with permission from [21].

side, and finally, (d) stabilizes the proton on the nitrogen side. The removal of the zero-frequency component from the optimal laser field reduces the isomerization efficiency by 9% and it reaches ~65%. The time variation of the isomerization probability is shown in Fig. 7.15.

References

[1] S.A. Rice and M. Zhao, *Optical Control of Molecular Dynamics*, John Wiley & Sons, 2000.

[2] P. Brumer and M. Shapiro, *Principles of the Quantum Control of Molecular Process*, John Wiley & Sons, 2003.

[3] R. Kosloff, S.A. Rice, P. Gaspard, S. Tersigni and D.J. Tannor, *Chem. Phys.* **139**, 201 (1989).

[4] S. Shi, A. Woody and H. Rabitz, *J. Chem. Phys.* **88**, 6870 (1988).

[5] C.D. Schwieters and H. Rabitz, *Phys. Rev.* **A48**, 2549 (1993).

[6] A. Kondorskiy, G. Mil'nikov and H. Nakamura, *Phys. Rev.* **A72**, 041401 (2005).

[7] A. Kondorskiy and H. Nakamura, *J. Theory Comput. Chem.* **4**, 75 (2005).

[8] A. Kondorskiy, G. Mil'nikov and H. Nakamura, Chapter 6 in *Progress in Ultrafast Intense Laser Science II*, Springer, 2007, pp. 119–142.

[9] M.F. Herman and E. Kluk, *Chem. Phys.* **91**, 27 (1984).

[10] E. Kluk, M.F. Herman and H.L. Davis, *J. Chem. Phys.* **84**, 326 (1986).

[11] M.F. Herman, *Annu. Rev. Phys. Chem.* **45**, 83 (1994).

[12] W.H. Miller, *J. Phys. Chem.* **105**, 2942 (2001).

[13] W.H. Miller, *Proc. Natl. Acad. Sci.* **102**, 6660 (2005); *J. Chem. Phys.* **125**, 132305 (2006).

[14] A. Walton and D. Manolopoulos, *Mol. Phys.* **87**, 961 (1996).

[15] E.J. Heller, *J. Chem. Phys.* **94**, 2723 (1998).

[16] H. Wang, X. Sun and W.H. Miller, *J. Chem. Phys.* **108**, 9726 (1998); X. Sun, H. Wang and W.H. Miller, *J. Chem. Phys.* **109**, 4190 (1998).

[17] M. Sugawara and Y. Fujimura, *J. Chem. Phys.* **100**, 5646 (1994).

[18] J. Manz and G.K. Paramonov, *J. Phys. Chem.* **97**, 12625 (1993).

[19] P. Mitric, M. Hartmann, J. Pittner, and V. Bonacic-Koutecky, *J. Phys. Chem.* **106**, 10477 (2002).

[20] H. Nakamura, *Adv. Chem. Phys.* **138**, 95 (2008).

[21] A. Kondorskiy, S. Nanbu, Y. Teranishi and H. Nakamura, *J. Phys. Chem.* **A114**, 6171 (2010).

[22] N. Elghobashi, P. Krause, J. Manz, and M. Oppel, *Phys. Chem. Chem. Phys.* **5**, 4806 (2003).

[23] H. Tamura, S. Nanbu, T. Ishida and H. Nakamura, *J. Chem. Phys.* **124**, 084313 (2006).
[24] H. Tamura, S. Nanbu, T. Ishida and H. Nakamura, *J. Chem. Phys.* **125**, 034307 (2006).
[25] W. Magnus, *Commun. Pure Appl. Math.* **VII**, 649 (1954).
[26] M. Irie, *Chem. Rev.* **100**, 1685 (2000).
[27] M. Irie, S. Kobayashi and M. Horie, *Science* **291**, 1769 (2000).
[28] M. Murakami, H. Miyasaka, T. Okada, S. Kobayashi and M. Irie, *J. Am. Chem. Soc.* **126**, 14764 (2004).
[29] K.L. Kompa and R.D. Levine, *Proc. Natl. Acad. Sci.* **16**, 410 (2001).
[30] A.H. Zewail, *Angew. Chem. Int. Ed.* **39**, 2586 (2000).
[31] W. Fuß, W.E. Schmid and S.A. Trushin, *J. Chem. Phys.* **112**, 8347 (2000).
[32] P. Celani, S. Ottani, M. Olivucci, F. Bernardi and M.A. Robb, *J. Am. Chem. Soc.* **116**, 10141 (1994); P. Celani, F. Bernardi, M.A. Robb and M. Olivucci, *J. Phys. Chem.* **100**, 19364 (1996); M. Garavelli, P. Celani, M. Fato, M.J. Bearpak, B.P. Smith, M. Olivucci and M.A. Robb, *J. Phys. Chem.* **A101**, 2023 (1997).
[33] M. Garavelli, C.S. Page, P. Celani, M. Olivucci, W.E. Schmid, S.A. Trushin and W. Fuß, *J. Phys. Chem.* **A105**, 4458 (2001).
[34] H. Tamura, S. Nanbu, T. Ishida and H. Nakamura, *Chem. Phys. Lett.* **401**, 487 (2005).
[35] A. Hofmann and R.de Vivie-Riedle, *J. Chem. Phys.* **112**, 5054 (2000).
[36] D. Geppert, L. Seyfath and R.de Vivie-Riedle, *Appl. Phys. B* **79**, 987 (2004).
[37] S. Nanbu, T. Ishida and H. Nakamura, *Chem. Sci.* **1**, 663 (2010).
[38] H.J.C. Jacobs and E. Havinga, *Photochemistry of Vitamin D and Its Isomers and of Simple Trienes*, John Wiley & Sons, 1979, Vol. 11, p. 305.
[39] S. Nanbu, M. Aoyagi, H. Kamisaki, H. Nakamura, W. Bian and K. Tanaka, *J. Theory Comput. Chem.* **1**, 263, 275, 285 (2002).
[40] A. Kondorskiy and H. Nakamura, *Phys. Rev. A* **77**, 043407 (2008).
[41] C.M. Dion, A. Keller, O. Atabek and A.D. Bandrauk, *Phys. Rev. A* **59**, 1382 (1999).
[42] W. Jakubetz and L.B. Leong, *Chem. Phys.* **217**, 375 (1997).
[43] J. Gong, A. Ma and S.A. Rice, *J. Chem. Phys.* **122**, 144311 (2005).
[44] T. van Mourik, G.J. Harris, O.L. Polyansky, J. Tennyson, J, A.G. Császár and P.J. Knowles, *J. Chem. Phys.* **115**, 3706 (2001).

Chapter 8

Enhancement and Suppression of Chemical Reactions by Continuous Wave Laser

By appropriately choosing the laser frequency and intensity of a continuous wave (CW) laser, it is possible to enhance reactions at low energies where the reaction probability of the original reaction is very small or suppress the reaction at high energies where the original reaction probability is high [1]. The time-independent nonadiabatic transitions in the Floquet or dressed state representation are used to design the reactions. The semiclassical analysis based on the Zhu-Nakamura theory of nonadiabatic transition can be carried out. Numerical demonstrations are made for two types of one-dimensional model potential systems. One is an ordinary-type barrier penetration reaction that cannot proceed efficiently at low energies. This reaction can be enhanced by dressing down an uncoupled excited state and creating the laser-induced curve crossings. At high energies where the reaction occurs efficiently, an appropriate choice of laser frequency and intensity suppresses the reaction. The second is an example of nonadiabatic chemical reactions of which the electron transfer is the most typical. Under the laser dressing, four coupled states are created and the reactions can be enhanced.

In reality, laser pulses are actually used. In this case it is good enough if the pulse width is wide enough to cover the reaction zone. The laser intensity required at least in the present model calculations ranges from 100 GW/cm^2 to a few TW/cm^2 and is weaker than the threshold (~10 TW/cm^2) to cause various undesirable multi-photon processes [2, 3].

8.1 Barrier Penetration Reaction

The first case we consider is a reaction which goes through the ground state potential energy surface that requires thermal activation because of potential barrier. The first excited state is assumed to be attractive, located a bit away from the ground state and not to affect the reaction along the ground state. The one-dimensional model potentials used are as follows:

$$V_1(x) = V_0 \mathrm{sech}^2(ax) \tag{8.1}$$

and

$$V_2(x) - \hbar\omega = b(x - x_0)^2 + c, \tag{8.2}$$

where $V_1(x)$ is chosen to mimic approximately the $H + H_2$ reaction with the parameters $V_0 = 0.457/27.21$ and $a = 1.36$ in atomic units [4]. The excited state $V_2(x)$ is chosen to be a shifted harmonic oscillator and the parameters used are $b = 0.01, x_0 = 0.3$, and $c = -0.003$ in atomic units. The original excited state potential $V_2(x)$ is assumed to be higher than the ground state by 0.07 a.u. This is actually not important, unless the excited state is too high and very high laser frequency is required to shift it down. Figure 8.1 shows the two adiabatic potentials obtained from Eqs. (8.1) and (8.2) with the coupling strength $V = 1.0 \times 10^{-3}$ a.u. which corresponds to the laser intensity $I \simeq 140$ GW/cm^2 in the case of transition dipole moment $\mu = 1.0$ a.u.

Numerically calculated transmission (reaction) probabilities are shown in Fig. 8.2 as a function of translational energy. The reduced mass of the system is taken to be 1060 a.u. which corresponds to the reduced mass of the $H + H_2$ system. The red curve is the reaction probability in the original potential case. The potential barrier height is around 0.015 a.u., as is seen from Fig. 8.1. The green curve with peaks and dips is the result under the CW laser. It is clear that we can enhance the reaction quite a bit at low energies where the original reaction probability is very small. On the other hand, the reaction is suppressed at around $E \sim 0.021$ a.u. where the original reaction probability is almost unity.

Figure 8.3 shows the reaction rate constant against temperature in the Arrhenius plot in the temperature range $300 \sim 1000$ K.

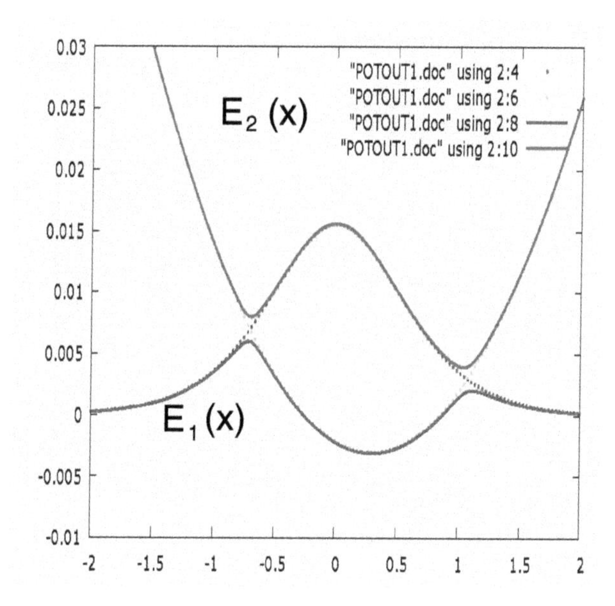

Figure 8.1: Adiabatic potentials $E_1(x)$ and $E_2(x)$ obtained from the diabatic ones given by $V_1(x)$ and $V_2(x)$ of Eqs. (8.1) and (8.2) with the laser-molecule interaction $V = 1.0 \times 10^{-3}$ *a.u.* Reproduced with permission from [1].

The enhancement is clearly seen. The widths of the probability enhancement peaks and suppression dips are dependent on the coupling strength V or the laser intensity and the potential shapes. Figure 8.4 shows the result for $V = 2.0 \times 10^{-3}$ a.u. The widths become slightly broader. In the case of simple one-crossing nonadiabatic tunneling-type potentials, the complete reflection dip becomes wider when the diabatic coupling is weaker [5]. When the system has more than one crossing like in the present case (Fig. 8.1), the dependence on the diabatic coupling is not monotonous. The peaks and dips depend also on the relative position of the potentials. Figure 8.5 shows the case of further shifting down the excited state $V_2(x)$ by 0.002 a.u.. The Fano-type resonance discussed by Vardi and Shapiro [6] under the name of laser catalysis can be reproduced by the present scheme which gives a general mechanism. The peak and dip at $E \simeq 0.0165$ a.u. in Fig. 8.2 is this type of resonance. At the other resonances the peak portion or the dip portion is enhanced.

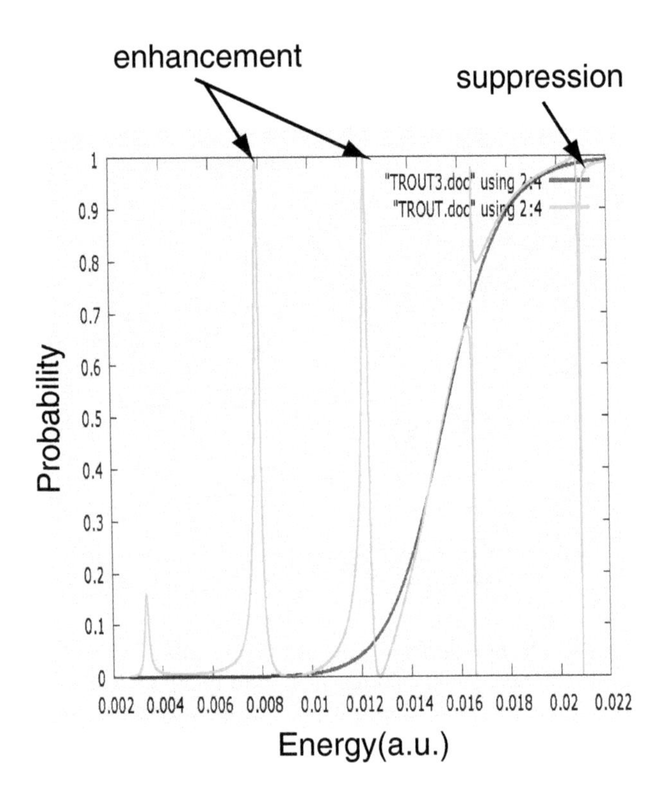

Figure 8.2: Reaction probability as a function of energy in the case of Fig. 8.1. The red curve represents the result of the original single potential case. The green curve indicates the effect of laser dressing. The three peaks at low energies represent the enhancement of the reaction. The dips at $E \sim 0.0125, 0.0165$ and 0.021 a.u. suppress the reaction. Reproduced with permission from [1].

The semiclassical analysis using the Zhu-Nakamura theory of nonadiabatic transition [5] can be done in order to have deeper understanding of the phenomenon. First of all, the corresponding diagram can be drawn as in Fig. 8.6. The arrows indicate the wave propagation from a certain position a to the other position b along the adiabatic potential n, the squares represent nonadiabatic transitions at avoided crossings and tunneling, and the circles with t indicate the turning points. The phase integral along the nth adiabatic potential is denoted as $\gamma_n(a, b)$. $I^\alpha(\alpha = R, L)$ is a matrix, representing the nonadiabatic transition at the right (R) or left (L) side of avoided

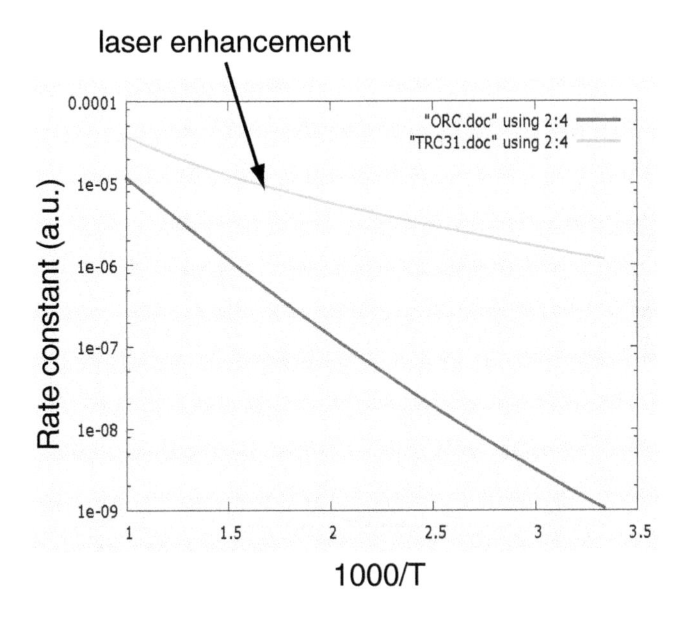

Figure 8.3: Arrhenius plot of reaction rate constant against temperature corresponding to Fig. 8.2. Reproduced with permission from [1].

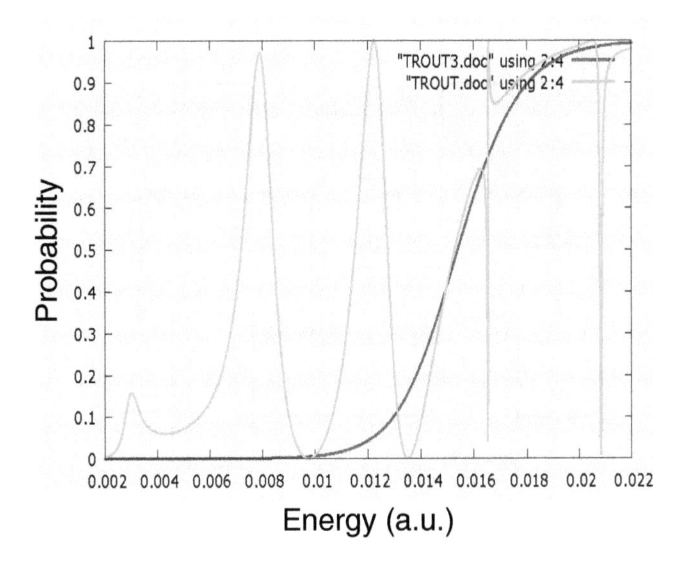

Figure 8.4: The same as Fig. 8.2 for the laser-molecule coupling strength $V = 2.0 \times 10^{-3}$ a.u. The peaks and dips are broader. The suppression occurs at $E \simeq 0.0135, 0.0165$, and 0.021 a.u. Reproduced with permission from [1].

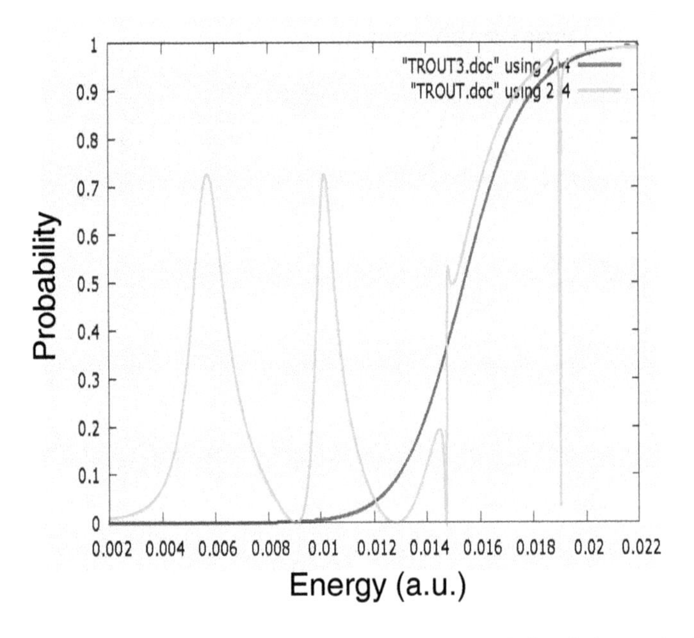

Figure 8.5: The same as Fig. 8.2 for a further downshift of the excited state potential from Fig. 8.1 by 0.002 a.u.. The laser-molecule coupling strength is the same as in Fig. 8.2 ($V = 1.0 \times 10^{-3}$ a.u.) Reproduced with permission from [1].

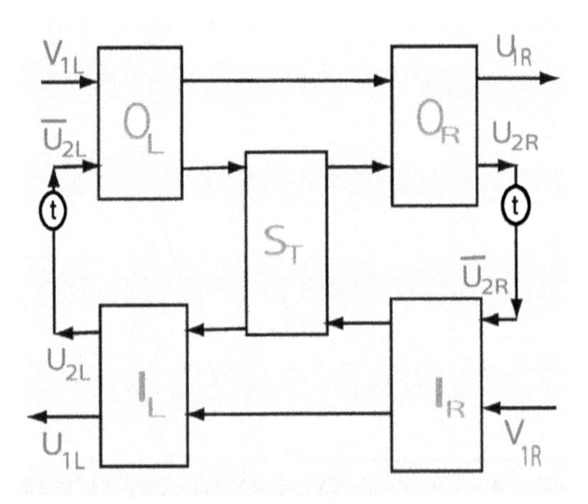

Figure 8.6: Diagram corresponding to Fig. 8.1 at $E < E_t$, where E_t is the barrier top of $E_2(x)$. Arrows denote the adiabatic wave propagation. The rectangles with matrices I and O represent nonadiabatic transitions at the right (R) and left (L) side of avoided crossing, and the rectangle with S_T represents the potential barrier penetration and reflection. Reproduced with permission from [1].

crossing. O^α is a transposed matrix of I^α. The element I_{21}, for instance, gives the transition amplitude from state 1 on the right side to state 2 on the left side. These are given in analytical forms as

$$\gamma_n(a,b) = \exp[i \int_a^b K_n(x)dx], \tag{8.3}$$

$$I^\alpha = \begin{pmatrix} \sqrt{1-p_\alpha}e^{i\phi_S^\alpha} & \sqrt{p_\alpha}e^{i\sigma_0^\alpha} \\ -\sqrt{p_\alpha}e^{-i\sigma_0^\alpha} & \sqrt{1-p_\alpha}e^{-i\phi_S^\alpha} \end{pmatrix}, \tag{8.4}$$

and

$$O^\alpha = \text{Transpose of } I^\alpha, \tag{8.5}$$

where

$$K_n(x) = \sqrt{\frac{2m}{\hbar^2}[E - E_n(x)]}. \tag{8.6}$$

p_α in Eq. (8.4) is the nonadiabatic transition probability for one passage of the avoided crossing point α. The various quantities including this p_α in Eq. (8.4) are provided by the Zhu-Nakamura theory [5].

S^T represents the transition matrix of barrier penetration. If we use the comparison equation method based on the Weber differential equation [7], we have

$$S^T = \begin{pmatrix} i\frac{e^{-\pi\epsilon}}{\sqrt{1+e^{-2\pi\epsilon}}}e^{-i\phi} & \frac{e^{-i\phi}}{\sqrt{1+e^{-2\pi\epsilon}}} \\ \frac{e^{-i\phi}}{\sqrt{1+e^{-2\pi\epsilon}}} & i\frac{e^{-\pi\epsilon}}{\sqrt{1+e^{-2\pi\epsilon}}}e^{-i\phi} \end{pmatrix}, \tag{8.7}$$

where

$$\epsilon = \begin{cases} \dfrac{-1}{\pi}\displaystyle\int_{T_L}^{T_R} |K_2(x)|dx & \text{for } E \le E_t \\ \text{Re}\left[\dfrac{1}{\pi i}\displaystyle\int_{T_L^*}^{T_R^*} K_2(x)dx\right] & \text{for } E \ge E_t, \end{cases} \tag{8.8}$$

and

$$\phi = \arg\Gamma\left(\frac{1}{2} + i\epsilon\right) - \epsilon\ln|\epsilon| + \epsilon. \tag{8.9}$$

T_L^* and T_R^* ($\text{Im}T_L^* < 0, \text{Im}T_R^* > 0$) are the complex turning points, and E_t represents the energy of the barrier top of $E_2(x)$.

Because of the quadratic potential barrier approximation to S^T, accurate quantitative reproduction of the numerical results cannot be obtained. The qualitative features of the whole process can, however, be analyzed as described below.

The labels attached to the arrows in Fig. 8.6 represent the coefficients of wave functions there propagating in the direction of arrow. The overall scattering matrix S can now be defined as

$$\begin{pmatrix} U_{1R} \\ U_{1L} \end{pmatrix} = S \begin{pmatrix} V_{1R} \\ V_{1L} \end{pmatrix}. \tag{8.10}$$

The adiabatic state $E_2(x)$ is an energetically closed channel as is depicted by the closed loop in the middle of Fig. 8.6. The whole dynamic process inside the loop is described by the following χ matrix:

$$\begin{pmatrix} U_{1R} \\ U_{1L} \\ U_{2R} \\ U_{2L} \end{pmatrix} = \chi \begin{pmatrix} V_{1R} \\ V_{1L} \\ \bar{U}_{2R} \\ \bar{U}_{2L} \end{pmatrix}. \tag{8.11}$$

The coefficients $U_{2\alpha}$ and $\bar{U}_{2\alpha}(\alpha = R, L)$ are connected by the adiabatic propagation through the turning point and are related as

$$\bar{U}_{2\alpha} = \gamma_2^2(x_s, x_l) \exp\left[i\frac{\pi}{2}\right] U_{2\alpha}, \tag{8.12}$$

where $x_{s(l)}$ is the smaller (larger) of the corresponding crossing point or turning point. The overall S matrix is now expressed in terms of χ matrix as

$$S = \chi_{OO} + \chi_{OC}[\Theta - \chi_{CC}]^{-1}\chi_{CO}, \tag{8.13}$$

where

$$\Theta_{nm}^{-1} = \delta_{nm}\gamma_2^2(x_s, x_l) \exp\left[i\frac{\pi}{2}\right] \tag{8.14}$$

and

$$\chi = \begin{pmatrix} \chi_{OO} & \chi_{OC} \\ \chi_{CO} & \chi_{CC} \end{pmatrix}, \tag{8.15}$$

where χ_{OC}, for instance, means the 2×2 submatrix, representing the open-closed portion of the 4×4 χ matrix. All the χ matrix elements are expressed in terms of the elements of the matrices I, O, and S^T. The phase $\exp[i\pi/2]$ in Eqs. (8.12) and (8.14) represents the effect of the turning point. All the elements of χ-matrix are explicitly given in terms of I, O and S^T matrix elements as shown in the Appendix. The diagonal (off-diagonal) elements of the S-matrix given by Eq. (8.13) represent the overall reflection (transmission) amplitude. The first term in this equation represents the direct process without trapping by the upper closed state and the second term represents the process trapped by this upper state. The destructive interference between these two waves at the exit to the left or at the exit to the right leads to complete reflection or complete transmission. Even if this interference is not complete, the reaction can be enhanced or suppressed due to this interference, as demonstrated above.

8.2 Nonadiabatic Tunneling-Type Reaction

Let us next consider a nonadiabatic tunneling type system of two coupled potentials. If we apply a CW laser and shift these potentials downward, we have four coupled potential system as shown in Fig. 8.7. The solid lines represent the original coupled diabatic potentials. These are defined as

$$V_1(x) = V_0[1 + \tanh(\alpha x)], \tag{8.16}$$

$$V_2(x) = V_0[1 + \tanh(\beta x)], \tag{8.17}$$

with the diabatic coupling

$$V_{12}(x) = V \exp[-\gamma x^2]. \tag{8.18}$$

These correspond approximately to the potentials along the reaction coordinate used in [8]. The parameters used are as follows in atomic units,

$$V_0 = 0.1, \quad \alpha = 0.8, \quad \beta = 0.9, V = 0.035, \quad \text{and} \quad \gamma = 0.28. \tag{8.19}$$

The potentials $V_3(x)$ and $V_4(x)$ are the laser-dressed states by the photon energy $\hbar\omega = 0.1$ a.u. and coupled by $V_{12}(x)$. The potentials $V_3(x)$ and $V_4(x)$ are coupled with $V_2(x)$ and $V_1(x)$, respectively, by

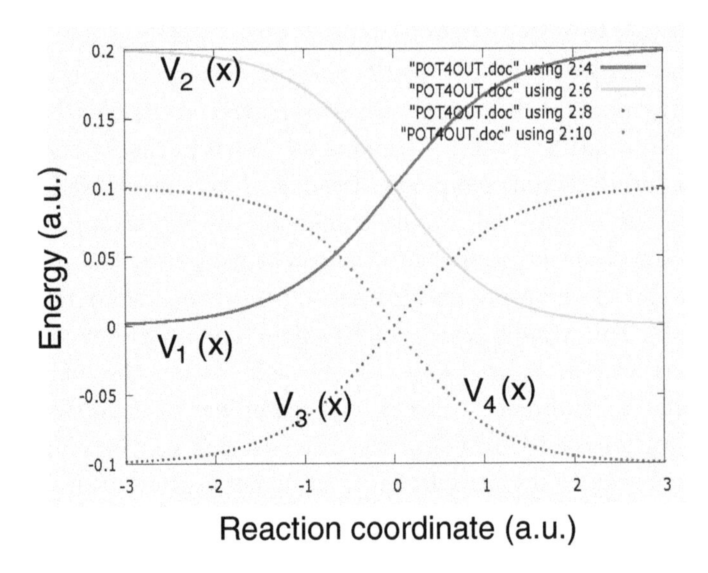

Figure 8.7: Four coupled diabatic potential curves. $V_1(x)$ and $V_2(x)$ are the original coupled diabatic states with the coupling V_{12}. $V_3(x) = V_1(x) - \hbar\omega$ and $V_4(x) = V_2(x) - \hbar\omega$ with $\hbar\omega = 0.1$ a.u. are the laser-dressed diabatic states with the same coupling V_{12}. $V_3(x)[V_4(x)]$ couples with $V_2(x)[V_1(x)]$ by the laser-molecule dipole interaction $V_{23} = V_{14} = 4.0 \times 10^{-3}$ a.u. Reproduced with permission from [1].

the dipole interaction $V_{23} = V_{14} = 4.0 \times 10^{-3}$ a.u. which corresponds to the laser intensity $I = 2.2\ \mathrm{TW/cm}^2$ in the case of transition dipole moment $\mu = 1.0$ a.u..

The numerical results of reaction probabilities as a function of energy are shown in Fig. 8.8. The red curve is the probability in the original coupled two-state case. The green curve indicates the enhancement by the laser. In the energy range shown here, the upper adiabatic states $E_3(x)$ and $E_4(x)$ (lower adiabatic states $E_1(x)$ and $E_2(x)$) $[E_1(x) < E_2(x) < E_3(x) < E_4(x)]$ are closed (open), and the reaction probability is the sum of the probabilities coming out along these open channels. The corresponding rate constants are shown in Fig. 8.9. The laser enhancement is clearly seen. The coupled two-state system like $V_1(x)$ and $V_2(x)$ considered here represents the basic one for electron transfer, namely Marcus's normal case [5, 9] and the present result suggests a possibility of its enhancement at

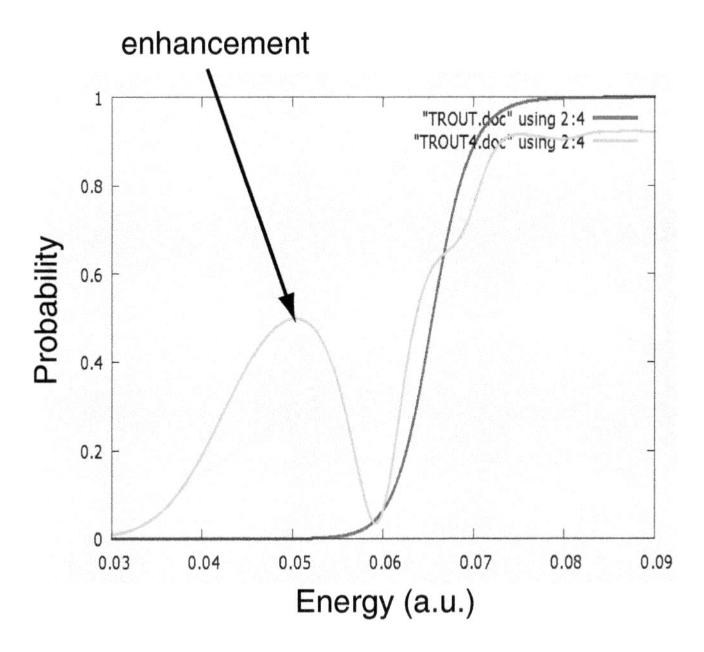

Figure 8.8: Reaction probabilities as a function of energy in the case of the potential system of Fig. 8.7. The red curve is the result in the case of the original coupled two states and the green curve shows the result of the four-state calculations, indicating the laser enhancement. Reproduced with permission from [1].

low temperatures. The complete reflection dips appear at higher energies at $E \gtrsim 0.14$ a.u. and can be used to suppress the reaction, if necessary. The enhancement of reaction probability at low energies is more interesting.

The semiclassical analysis can be performed in the same way as before. The diagram corresponding to the potential system of Fig. 8.7 is depicted in Fig. 8.10. The matrices I_C and $O_C (=$ transpose of $I_C)$ represent the nonadiabatic transition at the original crossing. The overall S-matrix and the χ-matrix are defined as follows:

$$\begin{pmatrix} \tilde{U}_{1R} \\ U_{1R} \\ \tilde{U}_{1L} \\ U_{1L} \end{pmatrix} = S \begin{pmatrix} \tilde{V}_{1R} \\ V_{1R} \\ \tilde{V}_{1L} \\ V_{1L} \end{pmatrix} \tag{8.20}$$

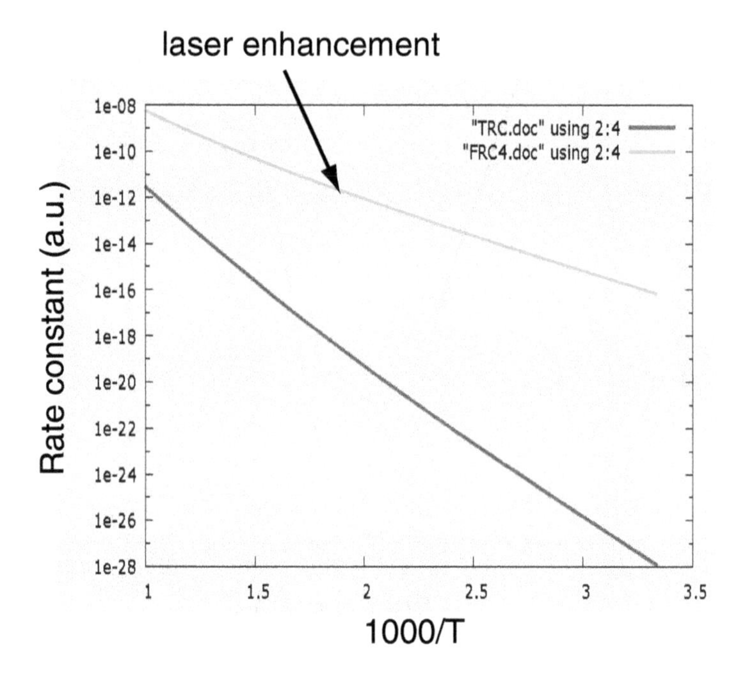

Figure 8.9: Reaction rate constant corresponding to the results in Fig. 8.8. Reproduced with permission from [1].

and

$$\begin{pmatrix} \tilde{U}_{1R} \\ U_{1R} \\ \tilde{U}_{1L} \\ U_{1L} \\ U_{2R} \\ U_{2L} \end{pmatrix} = \chi \begin{pmatrix} \tilde{V}_{1R} \\ V_{1R} \\ \tilde{V}_{1L} \\ V_{1L} \\ \bar{U}_{2R} \\ \bar{U}_{2L} \end{pmatrix}. \tag{8.21}$$

The expression of S-matrix in terms of χ-matrix is the same as Eq. (8.13). The matrix elements of χ can be obtained in the same way as before in terms of the elements of I, O and S^T elements.

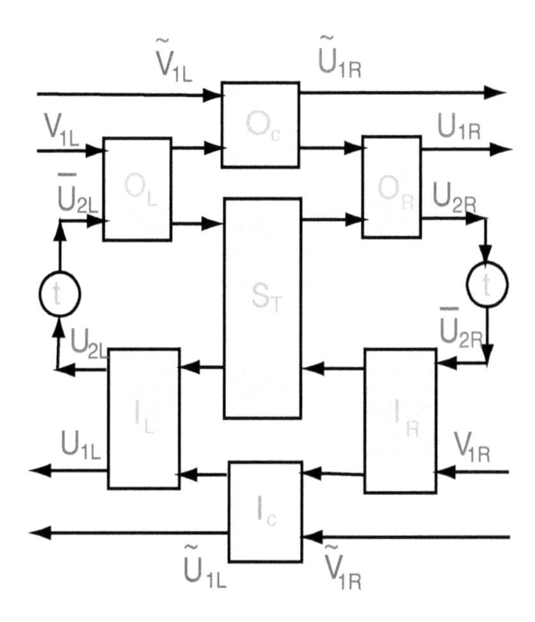

Figure 8.10: The diagram corresponding to Fig. 8.7. The meaning of each element is the same as that of Fig. 8.6. Reproduced with permission from [1].

Appendix

The matrix elements of χ defined by Eq. (8.13) is a symmetric matrix and can be explicitly expressed as follows:

$$\chi_{11} = O^R_{12} S^T_{11} I^R_{21} \gamma^2_{(2)}(r_{tn}, r_x), \tag{A.1}$$

$$\chi_{12} = O^R_{11} O^L_{11} \gamma_{(1)}(l_x, r_x) + O^R_{12} S^T_{12} O^L_{21} \gamma_{(2)}(r_{tn}, r_x) \gamma_{(2)}(l_x, l_{tn}), \tag{A.2}$$

$$\chi_{13} = O^R_{12} S^T_{11} I^R_{22} \gamma^2_{(2)}(r_{tn}, r_x), \tag{A.3}$$

$$\chi_{14} = O^R_{11} O^L_{12} \gamma_{(1)}(l_x, r_x) + O^R_{12} S^T_{12} O^L_{22} \gamma_{(2)}(r_{tn}, r_x) \gamma_{(2)}(l_x, l_{tn}), \tag{A.4}$$

$$\chi_{22} = I^L_{12} S^T_{22} O^L_{21} \gamma^2_{(2)}(l_x, l_{tn}), \tag{A.5}$$

$$\chi_{23} = I^L_{11} I^R_{12} \gamma_{(1)}(l_x, r_x) + I^L_{12} S^T_{21} I^R_{22} \gamma_{(2)}(r_{tn}, r_x) \gamma_{(2)}(l_x, l_{tn}), \tag{A.6}$$

$$\chi_{24} = I_{12}^{L} S_{22}^{T} O_{22}^{L} \gamma_{(2)}^{2}(l_x, l_{tn}), \tag{A.7}$$

$$\chi_{33} = O_{22}^{R} S_{11}^{T} I_{22}^{R} \gamma_{(2)}^{2}(r_{tn}, r_x), \tag{A.8}$$

$$\chi_{34} = O_{21}^{R} O_{12}^{L} \gamma_{(1)}(l_x, r_x) + O_{22}^{R} S_{12}^{T} O_{22}^{L} \gamma_{(2)}(r_{tn}, r_x) \gamma_{(2)}(l_x, l_{tn}), \tag{A.9}$$

and

$$\chi_{44} = I_{22}^{L} S_{22}^{T} O_{22}^{L} \gamma_{(2)}^{2}(l_x, l_{tn}), \tag{A.10}$$

where $l_x(r_x)$ and $l_{tn}(r_{tn})$ represent the left (right) side of the crossing point and the left (right) side of the turning point of the central potential barrier, respectively.

References

[1] Y. Teranishi, H. Nakamura and S.H. Lin, Chapter 5 in *Advances in Laser Physics and Technology*, Cambridge University Press, 2014, pp. 71–86.
[2] A.D. Bandrauk, Y. Fujimura and R.J. Gordon, *Laser Control and Manipulation of Molecules*, American Chemical Society, 2002.
[3] A. Kondorskiy, S. Nanbu, Y. Teranishi and H. Nakamura, *J. Phys. Chem.* **A114**, 6171 (2010).
[4] W.H. Miller, Y. Zhao, M. Ceotto and S. Yang, *J. Chem. Phys.* **119**, 1329 (2003).
[5] H. Nakamura, *Nonadiabatic Transition: Concepts, Basic Theories and Applications*, World Scientific, 2012 (2nd edition); *Introduction to Nonadiabatic Dynamics*, World Scientific, 2019.
[6] A. Vardi and M. Shapiro, *Phys. Rev.* **A58**, 1352 (1998).
[7] M.S. Child, *Semiclassical Mechanics with Molecular Applications*, Clarendon Press, 1991.
[8] C. Shin and S. Shin, *J. Chem. Phys.* **113**, 6528 (2000).
[9] R.A. Marcus and N. Sutin, *Biochim. Biophys. Acta* **811**, 265 (1985).

Chapter 9

Concluding Remarks and Future Perspectives

Controlling chemical reactions is not a simple dream anymore. Now we can challenge a variety of possibilities. This book has proposed various theoretical possibilities of controlling chemical dynamics by using lasers based on the achievements made by the author's research group. There are mainly three elementary processes in chemical dynamics: (1) excitation and de-excitation of energy levels or pump-dump of wave packets, (2) nonadiabatic transitions at conical intersections of molecular potential energy surfaces, and (3) motion of a wave packet on a single adiabatic potential energy surface. Any chemical dynamic process is a combination of these elementary processes and can be decomposed into a sequence of these elements. In other words, if we can control these elementary processes as we desire, the chemical dynamics as a whole can be controlled as we wish in principle.

One of the significant features of lasers is that the Floquet theorem or the dressed state picture holds well unless the frequency is very low. Energy levels or potential energy surfaces are dressed up or down by the amount of photon energy $\hbar\omega$. This indicates that new crossings, and thus new transitions, are created between the energy levels or potential energy surfaces. In the case of energy levels the crossings appear as a function of time-dependent laser frequency $\hbar\omega(t)$. In the case of molecular potential energy surfaces, new conical intersections are created. This means that there can be two types of conical intersections: one is the naturally existing intersection and the other is the laser-induced intersection. In any case this clearly implies that *nonadiabatic transitions*, both natural and induced ones, play crucial roles in controlling chemical dynamics [1, 2].

In the case of laser-induced intersections, we have proposed a new method of periodic sweeping of laser frequency at the newly created potential crossings. For the transitions among energy levels, the laser parameters can be designed so that the desired transition can be achieved selectively with 100% efficiency. The method is demonstrated numerically by taking (i) three and four closely lying energy levels, (ii) fine-structure energy levels of K and Cs atoms, and (iii) vibrational energy levels mimicking the ring-puckering isomerization of tri-methylenimine. The method can also be applied to excitation and photo-dissociation of molecules. Numerical applications are carried out for LiH and NaK diatomic molecules.

In the case of naturally existing conical intersections, the topography of the potential energy surfaces and the nonadiabatic coupling between the two surfaces are determined by nature and we cannot change them. It is true, in principle, to change the potential energy surface topography by applying a strong laser, but that strong laser easily induces many undesirable multi-photon processes and breaks up molecules eventually. It is estimated that it is better to use a laser with intensity less than $\sim$10 TW/cm^2 to avoid these undesirable processes [3]. Thus, in order to control the transitions at naturally existing conical intersections, it is required to change the momentum vector of the wave packet into the direction preferable for the transition in a multi-dimensional coordinate space. This can actually be realized by using the coordinate dependence of the dipole moment. This is called *directed momentum method.* More generally, the so-called *optimal control theory* has been designed by many authors [4]. It is to try to control the wave packet motion on a single adiabatic potential energy surface in the optimal way. Since the phases of both laser and molecular motion play crucial roles, the classical mechanical formulation, which can be done rather easily, is not good enough. The quantum mechanical version, on the other hand, is unfortunately too CPU-time consuming and not practical. In order to overcome this difficulty we have formulated a semiclassical version.

To improve the efficiency the appropriate intermediate states, i.e., the intermediate wave packets, are set on the way to the final target. This is called *semiclassical guided optimal control theory*, which has made it possible to deal with six-dimensional vibrational isomerization of the HCN molecule. This method, together with the directed momentum method, has also been applied successfully to photo-conversion of CHD (cyclohexadiene) to HT (hexatriene) and selective photo-dissociation of OHCl ($\rightarrow$ O + HCl). It is clearly demonstrated that the photo-conversion of CHD can be made with much higher efficiency than in the ordinary case. In the case of photo-dissociation of OHCl, the process, which cannot proceed usually because of the potential barrier, can be realized.

Another new possibility of using the intriguing *complete reflection* phenomenon has also been discussed. This phenomenon has been found in the nonadiabatic tunneling type of potential crossing, in which the transmission is stopped completely at certain discrete energies above the potential barrier top of the lower adiabatic potential. Three numerical demonstrations are made: one is the selective photo-dissociation of $\mathrm{HI}(^1\Sigma_+) \rightarrow \mathrm{H} + \mathrm{I}^{(*)}(^1\Pi_1, ^3\Pi_{0+}, ^3\Pi_1)$. By choosing the laser frequency properly, the dissociation product can be selected. The second and third are the photo-dissociations of two-dimensional models of $\mathrm{CH_3SH} \rightarrow \mathrm{CH_3} + \mathrm{SH}$ and HOD $\rightarrow$ H + OD or HO + D. In the first case the ordinarily favorable dissociation channel $\mathrm{CH_3S} + \mathrm{H}$ is stopped by choosing the laser frequency appropriately.

One more interesting possibility is enhancement and suppression of chemical reactions by a continuous wave laser. It is demonstrated that this is actually possible by appropriately choosing the laser frequency. One-dimensional models of a barrier penetration reaction and a nonadiabatic tunneling-type reaction are employed as examples.

In conclusion, four kinds of theoretical possibilities to control chemical dynamics have been proposed based on the achievements of the author's research group: (i) periodic sweeping of laser parameters, (ii) utilization of complete reflection phenomenon, (iii) guided optimal control theory and directed momentum method and (iv) enhancement and suppression of chemical reactions by a continuous wave laser. In principle, these methods can also be used to develop new molecular functions such as molecular switching and photo-conversion [1]. Naturally, however, further improvements and developments are needed. Furthermore, any experimental realizations are big challenges, but highly desirable. One advantage from the author's standpoint is that the formulations are based on a rather simple picture of dressed state and nonadiabatic transition. Only rather simple functionalities of the laser pulse envelope $\epsilon(t)$ and the laser frequency $\omega(t)$ are required. Collaborations and interplays between theory and experiment are inevitable.

References

[1] H. Nakamura, *Nonadiabatic Transition: Concepts, Basic Theories and Applications*, World Scientific, 2002 (1st edition), 2012 (2nd edition).
[2] H. Nakamura, *Introduction to Nonadiabatic Dynamics*, World Scientific, 2019.
[3] A. Bandrauk (Ed.), *Molecules in Laser Fields*, Marcel Dekker, 1994.
[4] S.A. Rice and M. Zhao, *Optical Control of Molecular Dynamics*, John Wiley & Sons, 2000.

Index